Sociobiology and the Preemption of Social Science

SOCIOBIOLOGY AND THE PREEMPTION OF SOCIAL SCIENCE

Alexander Rosenberg

THE JOHNS HOPKINS UNIVERSITY PRESS
Baltimore and London

This book has been brought to publication with the generous assistance of the Andrew W. Mellon Foundation.

The Johns Hopkins University Press, Baltimore, Maryland 21218
The Johns Hopkins Press Ltd., London

Library of Congress Cataloging in Publication Data

Rosenberg, Alexander, 1946-
    Sociobiology and the preemption of social science.
    Bibliography: pp. 219-21
    Includes index.
    1. Sociobiology.   2. Social sciences—Philosophy.
I. Title.
GN365.9.R66   306'.4   80-8091
ISBN 0-8018-2423-0

For S. T. Rosenberg and Blanca N. Rosenberg
  Because of their wisdom and in spite of their demurral

Contents

Preface

In the preface to a previous book, *Microeconomic Laws: A Philosophical Analysis,* I claimed that much important work in the philosophy of science is independent of grander issues in epistemology, and that the views defended in that book could neither be accepted nor rejected on the strength of alternative answers to fundamental questions in philosophy. This is a belief of which I have repented, and the present work is a reflection of that substantial change. While I still believe that findings about the actual practice of scientists can decide no fundamental metaphysical or epistemological issues, I now hold that the sides one takes on these issues must decisively determine the character of strategies of research in the natural and social sciences. The present essay offers a novel account of the implications of a commitment to empiricism for the research programs of the social sciences. It does so by pursuing an inductive argument to the best explanation of why the social sciences have failed to attain the degree and the kind of success in explanation and prediction that the natural sciences have attained, in spite of the employment of broadly similar empirical methods. The argument rests on the assumption that the methods in question, which reflect empiricist presuppositions, are as appropriate to the study of human behavior as to the study of any other natural phenomena, and therefore seeks the causes of failure in social science beyond alleged errors in method. This is the respect in which broad epistemological commitments shape narrower methodological decisions, and substantive contingent beliefs. For it is in such empirical beliefs that the mistakes and the failures of social science are to be found and explained. Or so this book shall argue. But its aims are not limited to the identification of false beliefs and the explanation of their firm grip on social scientists. For the diagnosis provides a prognosis, a prescription for improvement in these disciplines, which suggests that closer attention to methods and concepts drawn from the natural sciences, especially biology, will lead to successes where conventional social science has hitherto failed. The premises of my explanation of the failures of the social sciences are at the same time premises in an argument that they be replaced, superseded, pre-empted, by sociobiology.

So substantial a change of view as this work represents must be the pro-duct of very strong influences. Indeed, while no one hereafter acknowledged can be held to be responsible for or even in agreement with any of the claims broached, there are many persons with a very substantial causal role in the re-formation of my thinking. Fortunately for them, causal agency does not en-tail moral or intellectual responsibility for my radical conclusions. Earliest among the forces working to effect changes in my view were David Bray-brooke's persistent criticisms of my complacent treatment of economic theory. If my latest views are no more satisfactory to him than the earlier ones, I nevertheless owe him a great debt. I am equally indebted to my wife, Merle Kurzrock, whose equally persistent questions led me to first formulate the view about human behavior here defended, and led me to see its relevance for all the social sciences. As a biologist she also provided a spur and a resource for my thinking about the conceptual situation of biology. This is a subject unjustly neglected by the empiricist program in the philosophy of science, and one which, I was surprised to discover, is far more important to our understanding of social science than abstract prescriptions from the philosophy of physics. This book reflects that discovery, one I might not have made without my wife's influence.

For detailed comments and criticisms of material that eventuated in this work's account of social science, I am heavily indebted to Jonathan Bennett, Richmond Campbell, Eric Von Magnus, and Alan Donagan. For helpful dis-cussion of varying aspects of the issues here examined I owe thanks to Robert M. Martin, Daniel Hausman, Donald Davidson, Robert Cummins, Steven Strasnick, Laurence Davis, Alex Michalos, Alasdair MacIntyre, Joseph Pitt, Joseph Margolis, Jules Coleman, Ned McClennen, Stewart Thau, and Peter van Inwagen. Much of my thinking about biological theory which informs the later chapters of this book reflects the influence of Jaegwon Kim and invaluable discussions with David Hull, Michael Ruse, Mary Williams, Richard Burian, and William Wimsatt. For helping me to see how my apparently in-dependent interests in the social sciences and in biology come inevitably to-gether I must thank Thomas Lawson, Joan Straumanis, Phillip Scribner, Rada Dyson-Hudson, Werner Honig, Antony Kenny, Rom Harre, and Thomas Simon. For helpful comments on the exposition and assessment of their views I am also indebted to Gary Becker and E. O. Wilson. Although he is not cited in the pages below, Gilles-Gaston Granger's influence on my thoughts about the nature of science has been especially strong in this work. For reading and commenting on drafts of the entire manuscript I owe thanks to Robert J. Wolfson, Marshall Segall, David Hull, Michael Ruse, and especially Joel Kidder, who preserved me from infelicities and errors in every chapter. The large number of those that remain are but a mark of the still larger numbers from which he, and others, have preserved the reader. I must also thank Nita Esterline, Betsy Queen, and Val Cardoza for exceptional efficiency and thoroughness in transforming my original typescript into syntactically and orthographically readable copy. For further improvements in readability I am indebted to Carolyn Moser and Mary Lou Kenney. If this work is not yet readable, the author has only himself to blame.

For the opportunity to pursue this work I am indebted to several institutions: the Council for Philosophical Studies, at whose summer institute (sustained by the National Endowment for the Humanities) I shaped the earliest parts of this manuscript; the Maxwell School of Public Affairs at Syracuse University, which afforded me the opportunity to develop my views in the stimulating atmosphere of its interdisciplinary social science program; and finally, the University of California at Santa Cruz, whose invitation to serve as a visiting professor provided the freedom from interruption that enabled me to complete this project.

Finally, for bearing with several dislocations that permitted me to take advantage of these opportunities, I must thank Bloomsbury and the unselfish Gene.

Some of the material in this book appeared in different form in my "A Sceptical History of Microeconomic Theory," *Theory and Decision* 12 (1980): 79-93; "Can Economic Theory Explain Everything?" *Philosophy of the Social Sciences* 9 (1979): 509-29; "Obstacles to the Nomological Connection of Reasons and Actions," *Philosophy of the Social Sciences* 10 (1980): 79-91; "The Supervenience of Biological Concepts," *Philosophy of Science* 45 (1978): 368-86. The permission of the publishers and editors to use this material is gratefully acknowledged.

1

Introduction

The methods of investigation applicable to moral and social science must have been already described, if I have succeeded in enumerating and characterizing those of science in general. It remains, however, to examine which of those methods are more especially suited to the various branches of moral inquiry; under what peculiar facilities or difficulties they are there employed; how far the unsatisfactory state of those enquiries is owning to a wrong choice of methods, how far to want of skill in the application of the right ones; and what degree of ultimate success may be attained or hoped for by a better choice or more careful employment of logical processes appropriate to the case. In other words whether moral sciences exist, or can exist; to what degree of perfection they are susceptible of being carried; and by what selection or adaptation of ... methods... that degree of perfection is attainable.—John Stuart Mill, "On the Logic of the Moral Sciences," *A System of Logic*

It was with this statement that Mill launched the subject of the philosophy of social science as we know it. The assertion with which it begins and the questions that it poses remain the focus of both philosophers and social scientists eager to survey the nature and limits of our knowledge of human behavior. Mill's claim that the broadly empirical methods of natural science are fully appropriate to the aims of the human sciences has, in the century and more since he made it, been subject to a serious embarrassment. For in spite of widespread acceptance of Mill's methodological dicta in the social sciences, these subjects have remained as unsatisfactory as Mill found them. That is,

1

they have remained unsatisfactory by Mill's *standards*. For Mill went on from the passage above to write that "at the threshold of this enquiry we are met by an objection, which, if not removed, would be fatal to the attempt to treat human conduct as a subject of science. Are the actions of human beings, like other natural events, subject to invariable laws?"[1] But we seem no closer to such laws now, after several score more years of attempting to secure them, than Mill and his contemporaries were. The absence of such laws, or even of successively improved approximations to them, remains a continuing embarrassment to those empiricists who agree with Mill that the methods, and the sorts of knowledge which the application of such methods are to eventuate in, must be broadly the same in social and natural science. The absence of such laws was explained by Mill, and continues to be excused by his empiricist successors, on the grounds that the subject of the social sciences is "the most complex and most difficult subject of study on which the human mind can engage."[2]

This simple and rather obvious answer to the question has not won widespread acceptance, as the ever-burgeoning literature of the last hundred years' reflection on social science has shown. One reason philosophers have not been convinced by this answer is that it makes the differences between the two sorts of subjects a wholly contingent one, a difference only of degree by virtue only of the increased complexity of the subject matter of the social sciences. If Mill is correct, then the philosopher is deprived of a conceptual question that can provide grist for many mills.

More tellingly, the claim that the differences between social and natural science turn only on the increased complexity of the former's subject matter seems both too strong and too weak. Too strong, for it seriously underplays the vast complexity of many natural phenomena for which we now have fairly rigorous scientific accounts, in spite of their recalcitrance to observation or experimentation. Both the range and the degree of precision of explanation and prediction in the natural sciences have increased by several orders of magnitude since Mill wrote the *System of Logic*, and successive extensions of domains have made Mill's excuse for the relative backwardness of social science more and more hollow. Mill's appeal to complexity is too strong, for it suggests that with the increasing complexity of the subject matter of natural science from his own time to our own, progress in natural science should be decelerating and not accelerating. His explanation for the failure of social science is too weak because Mill credits the social sciences with already recognizing the concepts which are apt for describing and explaining human behavior in a scientific way. Human behavior, on Mill's view, is to be systematically explained by appeal to the same cluster of concepts— like motive, desire, belief, reason—that common sense appeals to in its explanation of actions. In fact he is committed to the same variables in the explanation of human activity that Plato embraced in the *Phaedo* over two millennia before him. By contrast, the chief obstacle that natural science has had to surmount is the absence of its explanatory vocabulary from common language, and the need to classify and describe phenomena in ways that cut across ordinary descriptions, in order to uncover general laws that regulate them. Thus Mill's explanation is too weak in light of the allegation that no significant progress has been made in the provision of a real *science* of man throughout the whole of recorded history. No progress has been made in spite of the fact that throughout the period we have been acquainted with the explanatory variables presumably required to generate this science.

The failure to explain plausibly the inability of social science to produce empirical laws has serious ramifications for an empiricist like Mill, who as-similates the aims and methods of the social sciences to those of the natural sciences. In the absence of a more plausible explanation he must either re-think his account of the methods of natural science or admit that there are important methodological and substantive differences between these two sorts of subjects. But embracing either of these alternatives involves the surrender of his views about the most central questions of philosophy. In either case the follower of Mill will have to surrender his commitment to the epistemological unity of all the methods of acquiring knowledge, and to the metaphysical unity of all the concrete subjects of systematic knowledge. Embracing either of these two alternatives is too sweeping a price to pay for an explanation of the failures of social science. Accordingly, in this work I hope to offer a new, more detailed and substantial explanation of the in-adequacies of social science, an explanation consistent with Mill's conviction that in describing the methods of natural science we thereby also describe those of social science. My explanation is thus subject to the constraints that it must admit the social sciences logically or conceptually capable of adopting the methods of natural science, and that it must allow for the formal possi-bility of these subjects' producing explanations and predictions of the same increasing degrees of precision that natural science provides. It must allow for these formal possibilities while more plausibly explaining the obvious con-tingent fact that the human sciences have failed in the employment of such methods to produce anything like such results. In short, my explanation, like Mill's, must be the explanation of what we both take to be a contingent state of affairs, a particular fact about the present state of our knowledge or lack of it about human behavior. Accordingly, my explanation, like Mill's, will hinge on contingent considerations themselves, albeit of the broadest possible sort. For no explanation of a contingent occurrence or fact can itself turn on only noncontingent logical necessities or conceptual truths.

The argument of the first half of this book has the structure of an infer-ence to the best explanation of the failures of the social sciences to attain the sort of success natural science has shown itself capable of. It is thus an in-ductive argument with a largely negative conclusion about the prospects for conventional social science. This pessimistic assessment differs from Mill's prognosis of eventual success because it turns on a stronger explanation of why these subjects have yet to produce scientifically respectable results. The work's aims are by no means limited to critical ones, however, for the explanation to be offered of the failures of social science provides a fairly specific guide for the direction in which an empirical science devoted to understanding human behavior should move. The argument does so by revealing an explicit and precise sense in which the social sciences must be *life sciences*, branches of biology, and can be expected to employ theories and make claims of no less and no greater generality than biological theory does. This positive result will be unexciting only to those with an antiquated or caricaturized impression of the theoretical scope and power of contemporary biology. Indeed if my argument is correct, the study of human behavior, conceived as a biological science, will admit of as much formally quantified and mathematical description as the most mathematical economist could hope for. As the second half of this book purports to show, in the end the explanation of the failures of contemporary social science is an argument for the success of *sociobiology*.

An important prerequisite of the argument to follow is some sort of prior attachment to the broad philosophical doctrines that motivated Mill and still sustain his empiricist followers. As noted above, the failure to find a plausible explanation for social science's inability to produce the desired results by the use of the prescribed methods forces on a follower of Mill the surrender of fundamental convictions in metaphysics and epistemology. *Mutatis mutandis*, the motivation to search for a plausible explanation here presupposes the same fundamental convictions. In the next chapter I attempt to trace out what these metaphysical and epistemological first principles are; to show that all the apparently methodological disputes in the philosophy of social sciences are but disguised variants of controversies about these broad metaphysical and epistemological principles; and to justify the positions attributed to Mill, and to empiricists generally, by showing that it is impossible to avoid taking sides on these philosophical issues, so that rejecting empiricist views commits one to embracing equally strong and, I believe, intrinsically less reasonable alternatives.

In addition to the constraints under which his philosophical commitment places the empiricist, his explanation of why the social sciences have failed to produce laws and theories of successively greater precision and accuracy is subject to a further condition. His explanation cannot do violence to one particular almost universal assumption about the determinants of human behavior. This is the assumption that these determinants are to be found in the joint operation of beliefs and desires of intentional agents. Mill clearly embraced this assumption, as his own discussion of the issue of free will reveals: "Given the motives which are present to an individual's mind, and given likewise the character and disposition of the individual, the manner in which he will act might be unerringly inferred; . . . if we knew the person thoroughly, and knew all the inducements which are acting upon him, we could foretell his conduct with as much certainty as we can predict any physical event."[3] This presumption that at least some and indeed most of our ordinary explanatory claims about particular actions, and the reasons which result in them, are true, is so well entrenched that some philosophers have argued that the denial of this claim is logically inconceivable.[4] In Chapter 3 the tenability of this further constraint on accounts of the failure of social science is explored. It is shown that even social scientists who purport explicitly to abjure this common assumption eventually find it unavoidable, and appeal to it tacitly in the very theories they claim to be free of it. Among these social scientists the most prominent are Durkheim and the structuralist followers of Lévi-Strauss. Examination of their views affords a fresh opportunity to reveal the epistemological consequences of denying the relevance of methods drawn from the natural sciences to the pursuit of social science.

Just as preserving this assumption is a constraint on the explanatory purposes of this work, the assumption that beliefs and desires explain the actions of human agents is itself a constraint on theories of human behavior. In Chapter 4 it is shown how this constraint, together with the methodological strictures of empiricism, directs the social scientist to search for *laws of human action*. Important episodes of this search are examined, and I argue that the modern history of economic theory in particular constitutes a sustained attempt to discover laws that will underwrite the singular claims made throughout the social sciences about particular reasons and their consequent actions. An examination of its history reveals that the theoretical shifts in economics are best understood as reflecting the failure to find laws of human action.

The conclusions of Chapter 4 pose a serious trilemma for the empiricist. He must render consistent (a) the common assumption that we are at least sometimes, indeed usually, correct in our specifications of particular desires and beliefs as the causes of particular actions; (b) the failure to find any law of human action to sustain this assumption; and (c) the empiricist view that causal claims must be sustained by laws. Mill's solution, of course, is to suggest that further work in social science can turn the failure that b reports into success. The objections to Mill's solution, however, demand a new way of circumventing the trilemma if empiricism is to retain its claim to our assent. Chapter 5 reveals the formal possibility of a path around the trilemma, through the examination of an apparently true, exceptionless general statement connecting reasons and actions in the way required for a law of human action. The candidate law is shown to be logically incapable of entrenchment in any system of other general statements, that is, in any scientific theory. This conclusion is exactly what would be expected if the explanatory factors cited in the candidate law did not designate causally homogeneous classes of events, states, and conditions, were not "natural kinds." If terms like 'desire', 'belief,' 'action', do not designate causally homogeneous classes of events, then they may indeed be used to express true singular causal statements, even though there is no law expressible in terms of these notions to sustain the singular claims. There will, on the empiricist's view, however, be other, unknown laws expressed in concepts hitherto and perhaps still unknown that will sustain the singular statements. The hypothesis that the terms we have hit upon to describe particular human actions and their causes do not reflect the features by virtue of which the two are causally connected enables the empiricist to render consistent the three horns of the trilemma he faces.

Nonetheless, the solution to the trilemma holds out what is at best a formal possibility. It lacks force because it has no grounds independent of its ability to outflank the difficulty empiricism faces. Chapter 6 provides this independent force for the solution, turning its merely formal possibility into material actuality. Independent reason is given to suppose that terms denoting reasons and actions are not natural-kind notions. Opponents of empirical methods in social science have long argued that these terms' meanings are given in connection with their exemplification by human beings, by members of the species *Homo sapiens*. In this chapter it is shown that biological theory requires that species' names be treated, not as kind-terms, but as proper names for spatiotemporally restricted particulars. Therefore, *Homo sapiens* is a name for a particular spatially distributed object and is not a purely qualitative predicate of the sort admissible in general laws. Since terms like 'desire', 'belief', 'action', etc., are to be defined through their semantical connection with '*Homo sapiens*', they cannot be purely qualitative predicates of the sort admissible in scientific laws and theories. The fact that laws relate natural kinds only thus excludes the possibility of laws of human action in a way that renders consistent empiricism and the truth of most of our singular judgments about particular reasons and their effects in behavior. Chapter 6 concludes by sketching the upshot of its argument about the spatiotemporally restricted character of species names for research programs in the operant conditioning of emitted behavior, the prospects for computer simulation of human activities, and philosophical controversies surrounding the notion of intentionality.

Much of the burden of Chapters 5 and 6 is borne by claims about details of contemporary biological theory. Thus, the failure of theoretical entrenchment for any exceptionless law of human action is established on analogy with the entrenchment of Mendelian laws of genetics. And the claim made in Chapter 6 about the character of species names hinges on understanding the actual character of the hierarchy of exception-ridden empirical generalizations, experimental laws protected by *ceteris paribus* clauses, and universal theoretical laws in evolutionary theory and in mathematical ecology. Important parallels and differences between these subjects, and work in social science employing formal techniques of precisely the same kind, are revealed and elaborated. This development is intended to show that the exclusion of species-related notions from the vocabulary of laws is not just a philosopher's trick to solve a philosopher's problem, but represents a fundamental constraint on scientific theory that nomologically successful subjects like biology have satisfied, and unsuccessful ones, like the social sciences, have not. Satisfying this constraint not only accounts for the success of biology, but will also enable us to discover those regularities that really do govern human behavior, to the extent that they obtain to be discovered.

In its attempt to substantiate this last claim, the focus of this work shifts from explanation of failure to prescription for success. In chapter 7 it is argued that, given practical limits, the empiricist demand that we search for laws of human behavior leads to biology, and in particular, to population biology as the locus of such laws. For these subjects, I argue, provide the narrowest natural kinds in which we can be confident that human behavior falls. This claim, in turn, is the very one which sociobiology needs to under-write its own claim to preempt all the conventional social sciences. Insofar as sociobiology requires this thesis for its fundamental rationale, empiricism is clearly wedded to this theory. But the empiricist has traditionally been skeptical about sociobiology's chief explanatory tool, the theory of natural selection. In order to allay suspicions that the theory is tautological or circu-lar, Chapter 7 closes with an analysis of its key theoretical term, 'fitness', and of the prospects for reduction of evolutionary theory to physical theory.

The power of sociobiology to curb unreasonable explanatory and pre-dictive expectations, and to provide powerful systematic accounts of human behavior is illustrated in Chapter 8. More important, misunderstandings of both sociobiologists and their critics about what is really crucial to the ac-ceptability of the theory are revealed in a critique of anthropological argu-ments against sociobiology. The claimed centrality of the problem of altruism to the acceptability of sociobiology is shown to be mistaken. The real issue between the sociobiologist and more conventional social scientists is the former's implicit claim to preempt the latter's discipline because sociobiology employs the narrowest natural kinds under which human behavior falls.

Yet insofar as sociobiological theory and its empiricist underpinnings in-volve denying the explanatory role of reasons and the appropriateness of labeling social explananda as types of actions, it seems open to the complaint that it is self-refuting and inconceivable. For to embrace the theory is to com-mit an action, to argue for it is to cite reasons as the causes of this action. In the last pages of this work I attempt to show that this charge of self-contra-diction misfires, although it does reveal the degree to which concepts in-appropriate for a science of man are entrenched in our ordinary views of him. We are thus presented with a forced choice between rejecting the relevance of this ordinary view for a science of human activity, or providing an alternative justification for the significance of a body of disciplines which cannot provide what the empiricist will accept as knowledge. Short of establishing the truth of the assumptions of empiricism directly, there is no firmer footing for the argument that follows than the choice this invitation provides to those who differ from empiricists in matters metaphysical and epistemological.

2

Metaphysics, Epistemology, and the Philosophy of Social Science

The most venerable and the most distinctive question raised in the philosophy of social sciences is also the most obvious one: Why are the sciences of man so different in their results from the natural sciences? In posing and answering this question, "different" is often taken as a euphemism for "inferior." This, indeed, is the way that this question will be understood in the present work, for its aim is to provide an answer to the question of why the social sciences have not made progress anything like that which has been made in the history of natural science.

The explanation to be offered begins with two assumptions of the broadest sort possible: I shall assume the truth of some version or another of epistemological empiricism, and of metaphysical materialism. In this chapter I justify such assumptions by showing that all the chief answers to the question about differences among natural and social sciences make such strong assumptions. Of course, we cannot hope to secure general assent to such assumptions within the philosophy of social science, and this subject can hardly wait upon the resolution of the grandest questions of the central concerns of philosophy. Therefore, the justification for my assumptions is nothing more or less than the claim that if my explanation for the failure of social science is a valid argument, anyone who disagrees with its conclusions and its imperatives for future work in the sciences of man is *ipso facto* committed to rationalism in epistemology, and to antimaterialism in metaphysics.

As noted in the first chapter, the question of why the social sciences are so different in their results from the natural sciences is the most venerable issue in the philosophy of these subjects because it was the starting point of the first sustained attempt to examine the methods and concepts of a science of man, Mill's *System of Logic*. Mill asks this very question within the context of certain presuppositions and with a distinct conviction about the degree and the kind of success that both natural and social science can attain. First of all, as the epigraph to Chapter 1 reveals, he assumed that both sorts of subjects are to be measured against the *same standards of success*. For he asks there "what *degree* of ultimate success may be attained" in the "moral sciences"[1] by the methods of natural science? This question, of course, reflects not only the presupposition that both fields should be assessed against the same standards of success, but also presuppositions about the character of natural science and thus about the methods by which success in the provision of a science is to be attained. Grandest of all Mill's presumptions is that the employment of the methods of natural science exhausts the routes to knowledge properly so called; that only by recourse to methods fundamentally scientific can we come to have rationally justified beliefs worthy of being characterized as knowledge. In effect, in raising and attempting to answer this question Mill took sides on at least three subsidiary questions in the philosophy of social science: (1) whether the social sciences *can* employ methods and concepts of the same kind as those of the natural sciences; (2) whether they *do* in fact employ such concepts and methods; (3) whether for purposes of acquiring knowledge of human behavior or action, they *should* employ such methods and concepts.

Mill's answers to these three questions were that (1) social science can do so, but that (2) to some extent social science fails to employ the right sort of methods and concepts for attaining its scientific purposes,[2] and that (3) it clearly should do so if it has any pretense to providing us with the sort of knowledge of human happenings that we want. His answer to the question of why social science has yet to attain the success of natural science was thus twofold: to some extent the failure lies in the use of unsuitable methods and "unscientific" concepts; more important, on Mill's view, the failure of the social sciences is attributable to the sheer complexity of their subjects—individual human agents and the social groups into which they aggregate. Individuals and social groups are just much more complicated than moving bodies, chemical reagents, and ocean tides. As I argued in the first chapter, this explanation both underplays the complexity of the subjects of natural science, and given the difference in the rates of progress evinced by natural and social science, overplays the recalcitrance to scientific treatment of individuals and social groups. For these and other reasons, criticism of Mill's explanation of the deficiencies of social science, and of Mill's answers to the questions of how social science can and should proceed, have provided most of the substance of the philosophy of social science from Mill's time to our own.

Much of this criticism reflects agreement with Mill that the social sciences have not yet produced laws and theories seriously deserving those names.

Mill's opponents have argued that this fact betokens failure only in the light of an inappropriate standard of assessment mistakenly imposed on social science by philosophers, misguided natural scientists, and even by confused social scientists. The social sciences have produced no laws, and no theories of the sort to be met with in natural science, just because their proper aims and methods have nothing to do with the provision of such results. What sterility and failure is to be found in social science, in this view, is largely due to the mistaken attempts of social scientists to employ methods suitable to the aims and subject matter of the natural sciences, but utterly inappropriate to those of social science. The social sciences are different from the natural sciences, but not inferior to them. Their differences are not in *degree* of success but in *kinds* of results. The two sorts of subjects are simply not comparable, and explanations like Mill's fail because they presume that the subjects are.[3] More often than not, Mill's opponents agree that at least some social science proceeds in accordance with the methods of natural science and employs concepts drawn from its repertoire, but they argue that the sterility and vacuity of these social sciences is due to their so doing. These opponents of Mill will also concur in the account of natural science that Mill and his followers have provided, but they will explain away the failure of social science to attain the successes of natural science by arguing that the apparent failure evaporates in the atmosphere of an appropriate standard.[4] This alternative to Mill thus provides a two-sided answer to the question, Why have the social sciences been so different in their results from the natural sciences? In part, the reason is because the aims and methods of social science are different from those of natural science, so that the latter's aims and methods can hardly be held up as a yardstick for measuring progress in the former; moreover, it is claimed, there has been great progress in the social sciences, progress to which philosophers like Mill have been blind because they have not looked for it, focused as their search has been on the provision of empirical laws and quantified theories. The second part of this explanation is the claim that to the extent that social scientists have taken up Mill's methods, the failure of their enterprise reflects the same misunderstanding of the aims of social science as that of Mill's philosophical followers.

Philosophers and social scientists who take this view of the questions about the status of social science answer them by pursuing a conceptual inquiry and applying its results to the diagnosis of social science gone *conceptually* astray. By contrast with Mill's contingent explanation of the failure of social science to attain the success of natural science, these thinkers offer an alternative conceptual analysis of the objectives and standards of social science, and purport to dissolve the question of its failures by showing that the question rests on conceptually unwarranted presumptions.[5]

The centrality of this disagreement to the pursuit of the philosophy of social science is no less obvious than the venerability of the question which generated it. But within the confines of this subject the dispute about the nature of social science is an irresolvable one. For the disagreement between Mill and his followers on the one side, and their opponents on the other, turns out to be one within the ambit of epistemology and metaphysics, and not at all, in the end, a parochial matter within the philosophy of social science. One side to the dispute claims that the nonexistence of social scientific results with both the generality and the practical applicability to prediction and control available in natural science reflects only the practical problems of complexity and difficulties of experimentation. Thus, these disputants are committed to a negative existential claim: there is no nonpractical obstacle to a natural science of man. Their opponents ridicule this view as the claim that the social sciences merely await their Galileo or Newton to synthesize the vast body of data already built up by empirical social scientists in expectation of the genius' appearance.[6] But these opponents are also committed to a negative existential claim: the claim that no such genius will appear, the claim that no laws and theories about human activity of the requisite level of generality and predictability will or can ever appear.

In general, there are two ways that a negative existential claim can be substantiated. One way to do so is to produce evidence of an unsuccessful search for the item whose existence is denied. This search must be exhaustive enough to sustain the negative claim in the eyes of reasonable men (as, for instance, our evidence that there is no Loch Ness monster or Abominable Snowman). The other method of justifying a negative existential claim is to deduce it from broader considerations to which one is already committed. Neither party to the present dispute will accept as sufficient the other's evidence of an open-minded, unsuccessful search for laws or conceptual obstacles to laws in social science. Moreover, no follower of Mill considers the history of failures to produce theories of human behavior to have been long enough or searching enough to establish even its practical impossibility. Nor will his opponent consider a mere catalogue of the vacuity and triviality of much "empirical" social research enough to sustain his thesis. It is thus to broader considerations of epistemology and metaphysics that disputants in the present case must appeal.

For Mill and his followers, the ultimate foundation of the diagnosis of the failures of social science reflects their commitment to versions of epistemological *empiricism* and/or metaphysical *materialism.*[7] Commitment to the first of these philosophical theses provides the justification for the claim that the standards of success imported from the natural sciences are appropriate for assessing the social sciences. For, if some version or other of empiricism is correct, then these are the only standards of cognitive adequacy that there are. Commitment to the second thesis—of materialism, or physicalism— provides the general argument to show that the social sciences are, at least in principle, capable of satisfying the standards drawn from the natural sciences, and that their failure to do so reflects at worst practical limitations or failures of industry and imaginativeness among social scientists. Of course, under both these labels lurks a variety of different epistemological and metaphysical theses, and it would not repay effort in this context to tease them apart from one another. All we need do is see how the most general characteristics of the variety of the philosophical claims that go under these names have the consequences described.

Empiricism is a label for those theories according to which knowledge is ultimately justified by observation, by experiment, by experience, by some sort of sensory awareness of happenings that are in one way or another not under our cognitive control. Of course, throughout the history of philosophy no one has been able to spell out a version of this thesis free from difficulties, but this has not deterred philosophers from embracing the claim that at least some such thesis is correct, even if they cannot yet state it correctly. Some versions of the empiricist's thesis also allow for the truth of statements that are not justified by experience, such as those of mathematics, but only on the ground that such statements are conventional (again in ways that have not yet found entirely adequate expression). In any case, such statements can have empirically significant bearing on our knowledge only to the extent that experience determines that their subject-terms are realized among the happenings beyond our cognitive control. Thus, Euclidean geometry may be treated as a body of nonempirically justified necessary truths, but then the question arises whether there actually are any Euclidean triangles or not. For practical application, a body of nonempirical truths must be conjoined to existence claims about their subjects; and the conjunction of these two sets of statements is, on the empiricist's view, as much in need of experiential justifications as any other claims to knowledge are. Of course, any claim to knowledge that transcends past or present justificatory evidence requires test against future happenings beyond our cognitive control. Since these strictures on justification are, on the empiricist's view, exhaustive, it follows that any claim to knowledge must satisfy them. If, as the empiricist believes, the findings of natural science and the methods that eventuate in these findings satisfy these adequacy conditions, then the question of whether the social sciences can satisfy the standards met by the natural sciences, broadly conceived, is identical to the question of whether they can provide knowledge at all. Since it is agreed by all parties that the social sciences should be expected to provide *knowledge*, it follows, according to the empiricist, that these subjects *should* employ the methods and concepts of the natural sciences, broadly conceived.

The conviction that the social sciences *can* successfully employ these methods and concepts turns on embracing another set of beliefs, those which we may conveniently label physicalism (materialism is a more traditional name for this view, but since it has been preempted in the philosophy of social sciences by Marxians, I shall avoid it). In its crudest form, physicalism has it that human beings, the subjects of the social sciences, are nothing but aggregations of the items with which the natural sciences deal. We are composed exclusively out of materials dealt with by theories in biology, chemistry, and physics. Furthermore, the physicalist embraces other theses that can be given fancy labels: he is a methodological individualist and a mereological determinist. That is, he holds that aggregations of human beings and their behavior are exclusively composed of and determined by their component individuals and these individuals' behavior; and that the individual human agents are also exclusively composed of physical substances and their behavior is exclusively determined by the behavior of these component parts. As a consequence of his commitment to these doctrines, the physicalist is committed to the theory that the behavior of groups and individuals is fully determined by the behavior of the physical constituents of human beings. If the behavior of these constituents is determined in accordance with the operation of general laws, then it follows that human behavior must also be governed by these laws and by whatever generalizations about human beings follow from these laws. Accordingly, (1) if the natural sciences can uncover laws governing the behavior of their subjects, and (2) if the subjects of the social sciences are composed exclusively out of the subjects of the natural sciences, and (3) if their aggregate behavior is exclusively determined by laws governing their individual behavior, it follows that (4) the social sciences can, at least in principle, discover laws of human behavior—for these laws will either be the laws governing the constituents of human beings, or laws that can be logically derived from these more fundamental generalizations. Since on the assumptions of physicalism, laws of social science are at least in principle possible, it follows that their absence is no reflection on their impossibility.

It must be admitted that, historically speaking, few followers of Mill have explicitly embraced views like these. Indeed, while Mill did claim that the behavior of aggregations of human beings was to be explained exclusively by appeal to the behavior of the individuals, he remained agnostic about the possibility of explaining individual behavior exclusively in terms of the behavior of component physical parts of the individual.[8] There are a number of reasons why Mill's followers have not publicly offered such views as part of their explanation for the possibility of a science of human behavior. One reason is the philosophically controversial character of these views, which unsuits them for convincing any but the already converted. Another is the degree to which the claims of mereological (part-whole) determinism seem to rest on the pursuit of a research program in neurology, anatomy, and physiology whose eventual success seems clearly an empirical matter. Moreover, conviction with regard to the claims of physicalism is not required in order to guide the methods of the social sciences; the commitment to empiricism already assures us that *if any method will provide knowledge* of human behavior it is the method of natural sciences, and that the expression of this knowledge will employ concepts of the same general kind as those which figure in the latter disciplines. Thus, further argument of a controversial kind to the effect that the social sciences can succeed by the use of these methods is not only superfluous, but may generate argumentative digressions from the empiricist's analysis of and prescriptions for the methods and concepts of social science. Nevertheless, physicalism does provide a *positive* argument for the possibility of a scientific treatment of human behavior. It provides an implicit guarantee that the attempt to construct such a scientific discipline, motivated by empiricist convictions, is not only a necessary condition for knowledge of human behavior but, given enough industry and ingenuity, sufficient as well.

It should now be clear why the follower of Mill answers in the affirmative to the question of whether the social sciences should and can proceed in accordance with the methods of natural sciences. It should be equally clear that, if he is a physicalist, his only answer to the question of why they have not attained the same level of success must be something like the conclusion that the subject matter of the social sciences has proved hitherto more refractory and complex than the subjects of the natural sciences. He is committed to his answers to these questions by considerations that far transcend the traditional confines of the philosophy of social science.

But by the same token, the modal force of the alternative views—that social sciences, logically, neither can nor should proceed in accordance with the account of science offered by Mill or his successors—presupposes philosophical commitments of at least as grand a scale as empiricism and physicalism. For minimally, they involve the denial of any version of empiricism or physicalism strong enough to underwrite the possibility of a social science in the image of natural science. If a philosopher argues that the epistemic standards imposed by empiricist considerations are inappropriate to assess the methods and results of social science, then he must either deny that these subjects provide knowledge properly so called, or he allows for the possibility of nonempiricist justification for the findings of social science. If he goes on to claim that significant, explanatory, nonempirically justified claims about human activities are possible, indeed actual, where empirically justified ones are not, then he must embrace some alternative to the empiricist's account of the nature, extent, and justification of knowledge. Of course, some opponents of empiricism are avowed rationalists[9] (for this is what denial of the empiricist's claims comes to). But most philosophers who reject the claims of Mill and his followers do not recognize or admit these labels, and do not provide the epistemological impedimenta that their denials require. They fail to do so in some cases simply because they do not recognize the modal force of their claims, that their denials have the force of logical necessity.[10] Others fail to do so because they do not grant the suitability of these labels, supposing that issues in the philosophy of science are independent of matters metaphysical and epistemological.[11] Or finally, some believe philosophy to have transcended the issues on whose existence the intelligibility of these labels hinges.[12]

That Mill's opponents are nevertheless committed to rationalist and anti-materialist doctrines willy-nilly becomes clear in their arguments for the claim that the social sciences meet different standards from those empiricists have established or detected in natural science, and do so by the employment of concepts incommensurable with those of natural science. For their principal arguments invariably conclude that the explanatory variables in social science cannot be physical states of individual agents, nor are they even causes of the happenings they are claimed to explain.[13] Both followers and opponents of Mill have agreed with his view that distinctively human activities are correctly explained by the citation of reasons, desires, beliefs, motives, expectations, hopes, fears, wants, aims, purposes, intentions. Accordingly, a demonstration that these items could not be physical states of individuals would be tantamount to a refutation of physicalism. The rejection of physicalism, as noted above, is not in and of itself incompatible with the view that the social sciences *ought* to pursue the methods of natural science. Physicalism is only a sufficient, and not a necessary, condition for the employment of these methods. Mill's opponents, however, argue that social science does not, and cannot, satisfy standards imposed from natural science, but that it *ought not* do so either. And this imperative hinges on the denial that social inquiry is *causal* inquiry,[14] which in turn commits its exponents to the denial of empiricism as a theory of knowledge.

On these philosopher's views, the aim of the social sciences is to produce an understanding of social life different in kind from that which natural science produces for natural phenomena. But if this understanding is to have the standing of *knowledge* of social phenomena, then either it must have a justificatory foundation acceptable to empiricist epistemology, or that epistemology must surrender its pretensions to exclusiveness and exhaustiveness. The denial that the understanding of social life provides causal discoveries is in effect the denial of this very pretension of empiricism. For by 'causation' both the proponents and opponents of Mill's view of the sciences mean a relation whose character, and all knowledge of which, turns on embracing the empiricist's account of knowledge and standards for cognitive success. The account of causation to which almost all parties to the dispute about the differences between natural and social sciences agree is the so-called Humean or regularity theory. According to this theory, causal relations obtain between items by virtue of their subsumption under contingent general regularities or laws. But our knowledge of these regularities can only be justified empirically. The empiricist, following Hume, fails to find any singly detectable property common and peculiar to all causes and all effects, except the property of being members of classes of events constantly conjoined. And if these detectable properties represent the limits of knowledge, then it follows that our knowledge of causal relations must consist in acquaintance with empirical regularities, and the justification of these regularities must involve their employment in prediction and control. The empiricist is constrained to deny, for want of any evidence, the existence of a stronger link between cause and effect than that of *de facto* regularity; in particular, he must deny the role of causal agency, power, or any other *necessary connection* between individual causes and their effects.

It is because natural science involves the search for causes, and *ipso facto*, the formulation of laws expressing empirical regularities that it manifests the exigencies of empiricist epistemology. Equally, any subject that eschews the search for causes thus construed eschews the search for laws as well. But without laws there is no prospect, on the empiricist view, of systematic and justifiable prediction and control of future occurrences, and accordingly no scope for testing the truth of the findings of a science.

If, in the absence of laws or their approximations, a body of propositions can still lay claim to providing understanding, to constituting knowledge of social phenomena, it can do so only on the assumption that empiricism is not the exclusive and exhaustive epistemology its proponents take it to be. In particular, the justification of the general truths that a nonempirical discipline employs to explain the phenomena with which it deals will perforce demand an epistemological alternative to empiricism; and the anti-empiricist philosopher of social science is willy-nilly compelled to provide one. Thus, if he denies that the reasons we cite in explaining particular actions are causes of these actions, on the ground that there are no empirical lawful generalizations possible linking reasons and actions, then on pain of surrendering the legitimacy of all such explanations, the philosopher must offer us an account of how these reasons do explain their consequent actions. If they do so because of some logical connection between the reasons and the actions, then the philosopher owes an account of how logical connections can explain contingent occurrences. If reasons explain actions because of some sort of noncontingent but nonlogical connections, then our knowledge of such connections must be accounted for. Since only the possibility of a priori knowledge of synthetic truths could do this, this latter alternative commits Mill's opponents to rationalism. It is largely because anti-empiricists have not recognized that they owe answers to these questions that they have not realized that they are committed to the direct denial of empiricism and to the support of rationalism as the theory of knowledge on which their conception of social science must be founded. Philosophers who have recognized the force of these questions have either embraced rationalism as a philosophy of social science, or embraced rationalism as a philosophy of science *tout court*. But antiempiricist philosophers who are conscious of their obligation to propound or defend an epistemological alternative are the exception.

Of course to the extent that philosophers of social science consider themselves to be answering only the question of whether the social sciences actually do employ the methods and concepts of the natural sciences, they seem to be under no obligation to embrace any particular grander claim in either epistemology or metaphysics. For any description of how these subjects actually do proceed is quite neutral as between various assessments of their success and alternative commitments to their cognitive merits. One could accept the view that the social sciences, as actually pursued, are different in kind from the natural sciences, while still holding the view that they ought not to be, and that until their methods and concepts are reconstituted on a properly "scientific" footing, they have no more claim on our credulity than phrenology or astrology. Indeed, many contemporary philosophers claim to be doing no more than giving an account of the actual conceptual situation of the subjects which they investigate; and they consider questions of how these sciences should proceed, or indeed can proceed, legislative prescriptions beyond their rights or competence. But this fine impartiality is not only rare, it is also hollow and unconvincing. For all parties to the dispute agree that they are examining disciplines which ought to provide knowledge of human affairs, and their attacks on alternative accounts of the situation of the social sciences inevitably reflect implicit commitments to grander views in philosophy just because all parties treat these subjects as something more than demonology or alchemy. Whether the social sciences can or cannot, should or should not, be measured by standards imported from the natural sciences makes a staggeringly important difference to practical affairs, to the way individuals and groups are treated, not to mention manipulated. Neutrality on questions of metaphysics and epistemology is quite untenable in the face of the venerable question with which the modern subject of the philosophy of social sciences began: Why are the social sciences so different in their *degree* or their *kind* of success from the natural sciences? This chief topic of puzzlement in our discipline presupposes that there is a difference; it requires that we take sides on whether this difference is one of degree or kind; and it calls for an explanation of the difference. The social sciences have as their aim to provide knowledge, and they purport to do so. Taking the legitimacy of their methods and concepts at face value means taking their aims and claims at face value as well. The philosopher who does the former can hardly refrain from doing the latter. Taking their aims at face value and admitting that these subjects have not attained the same degree of success as others entails assessing them against the same standards as the others, and it means accepting some version or other of empiricism and perhaps also physicalism. Taking the aims and claims of social science seriously, but asserting that they have succeeded, albeit in the light of different kinds of standards, *mutatis mutandis* entails some version or other of rationalism and perhaps also of antimaterialism.

It is easy to see that all the questions that are the philosophy of social science's stock-in-trade ultimately depend either for their motivation or for their answers on how we account for the difference between the natural and the social sciences. Consequently, the entire subject in effect awaits the answers to those traditional questions in philosophy that many (including the present writer) thought they were escaping from.[15] This seems perhaps most obvious in the problem of the nature of explanation of human action in history and the other social sciences. The empiricist insists on the existence of covering laws in the absence of any apparent candidates and in the face of the historian's denials. This insistence reflects his conviction that the only cognitively respectable explanations of particular happenings are causal explanations and *ipso facto* presuppose the existence, if not the knowledge, of laws. On the other hand, those who, citing and approving actual practice of historians, deny the existence of such laws and deny the causal character of the historian's explanations (or opt for a nonnomological sort of causation) can sustain their convictions of cognitive respectability only on footing no less substantial than that of their opponents. It seems just as true, if slightly less obvious, that most of the other traditional questions in the philosophy of social sciences hinge on these same themes: the conviction that functional, purposive, or teleological explanations can ultimately be given an analysis in terms of mechanical or Humean causal explanation and the opposite commitment that, although intelligible and legitimate forms, these sorts of explanation are *sui generis*, must, in the end, be decided in the context of the wider questions of philosophy. Equally, the dispute between holists and individualists like Durkheim and Popper is unlikely ever to be settled in the alternation of proffered reductions of societal concepts to individual ones and their rejection on grounds of counterexample and circularity. And again, this is because the parties to the dispute have much more fundamental differences over physicalism and holism. Or again, the varying disputes about the actuality or possibility of "value-free" social sciences reflect differences about aims and standards of the social disciplines, aims dictated by alternative conceptions of what sort of understanding these subjects do and should provide. Even apparently arcane disputes, like that over the possibility of and the criteria for the simulation of human action by computers, in the end hinge on these same broad issues in epistemology and metaphysics.

It would be the depths of unreasonableness to expect resolution of the main problems of philosophy; indeed, to expect solutions to these problems may well constitute the surest sign that they are misunderstood. But it would be only slightly less unreasonable to expect that work in the philosophy of the social sciences should therefore cease. This is not just because the subject has its own academic momentum, quite apart from any practical significance; more important, philosophers who have taken sides, rightly or wrongly, on the grand questions of epistemology and metaphysics are in a position to expound the consequences of the sides they have chosen for the questions of how social science should be pursued, and why it is so different from natural science. The trouble is they cannot expect a fair hearing, let alone agreement, from those who have taken the other side of these vast issues. Nevertheless, they can hope to influence actual practitioners in the social sciences who have taken their side in matters philosophical, and want to know whether and what implications for their work follow from the sides they have chosen. Thus we should hardly demand that work under the title "philosophy of social science" cease, pending the solution of the main problems of philosophy, *if* epistemologically and metaphysically committed philosophers come to believe that the implications of their commitments have not yet been recognized or fully appreciated by social scientists, and *if* they further believe that such recognition and appreciation will have a material and beneficial effect on the pursuit of social science. Such a view often goes hand in hand with an explanation of why those aspects of social research not hitherto animated by the alleged discoveries of the committed philosopher have been sterile failures. In fact, such a view is often the result of the committed philosopher's reflection on the differences between the natural and social sciences. In sum, the philosopher who pursues these matters is clearly not taking up "open questions" in a distinct and relatively autonomous subdiscipline; rather, his work is more like exegesis or casuistry, working out the application of received and unquestioned first principles, shedding light on them and with them.

This is precisely what I propose to do in this work. For I am one of these committed philosophers, and I believe that the full consequences of the truth of the grand theses that I am committed to have not yet been completely explored, nor has their beneficial influence been sufficiently applied to the advantage of progress in the social sciences. The views to which I am committed are versions of empiricism in epistemology and physicalism in metaphysics earlier attributed to followers of Mill. Such commitments, I argued in the first chapter, seem to constrain my explanation of the relative failure of social science to produce results nearly as impressive as the natural sciences. They constrain the explanation to one in terms of mere practical complexity and the want of sufficient industry and imagination. But this is a view I stigmatized as implausible because it underestimates the vast complexity of subjects in which the natural sciences have made substantial progress, and seems too weak to explain the failure of an attempt to erect a science of man that employs concepts with which we have been describing our own behavior in all recorded history. Accordingly, my commitment to empiricism and physicalism obliges me either to find a better explanation for why the social sciences have not attained a standing that on my views must be possible for them, or to explain why the appeal to the practical obstacle of complexity is satisfactory even in the face of this implausibility. This, indeed, is the obligation the present work proposes to discharge: to provide a new explanation of why the social sciences, though conceptually and physically capable of employing the methods and concepts of natural science, and capable of providing explanations and predictions of the same increasing degrees of precision as the natural sciences, have hitherto failed to do so. This explanation trades on considerations akin to the complexity that Mill appealed to, and is clearly "empirical" and not "conceptual" in character. Nevertheless, it is sufficiently stronger than Mill's not to fall prey to the implausibility that haunts his view of the matter. It is stronger because it hinges on just those two considerations that render Mill's explanation implausible: the complexity of natural phenomena, and the character of the concepts ubiquitous throughout social science and ordinary life. The explanation to follow trades on the manner in which biology treats highly complex natural phenomena and on the factual (though not logical) irreducibility of these concepts of ordinary life and social science to those of biology.

It is a long-standing tradition among philosophers and social scientists to find fault with social sciences as their contemporaries pursue it and to recommend a radical revision of the methods and concepts in use along the lines of or on analogy with some other favored science which the writer approves of as providing real knowledge. More often than not, physics is the science held up for backward disciplines to emulate. And much of the history of changes in the methods of social science reflects changes in the perceived character of physics and its methods and concepts: atomism, Newtonian determinism, field theories, relativity, quantum mechanical uncertainty, and complementarity have all been used and abused in the service of one revisionary proposal or another offered to revolutionize social science and set it definitively on the high road to scientific respectability. But except in the works of Herbert Spencer and some sociological functionalists, biological thinking has hardly been appealed to at all as the model on which a social science should be constructed. This is as much a reflection on the state of biological theory throughout the period leading to the last thirty years as it is a reflection on the glorious history of physics in the last three hundred years. But recent advances in fields like quantitative biology make this science a more and more suitable conceptual model for the more "backward" disciplines. Moreover, if the physicalist is correct, the social sciences must be pursued not just on a methodological analogy with the life sciences but as part and parcel of them. For, if the physicalist is correct, there is nothing more to the subjects of the social sciences than their status as living organisms, complex aggregations of just the material that biology treats. Accordingly, it will be an especially strong confirmation of the empiricist's explanation of the failures of current social science if this explanation hinges on the assumption that human beings are nothing but biological organisms whose behavior can be explained only up to whatever limits are set on biological explanations of the behavior of organisms in general. Furthermore, such a finding will provide all the force that stands behind empiricism and physicalism as a motivation or rationale for an entire program of research in social science, a program about whose eventual success there can be no more doubt than is reasonable to maintain about the success of biology itself. It is in this sense that I claim the full implications of empiricism and physicalism have not yet been realized in the social sciences, and that recognizing and appreciating them will have material and beneficial consequences for the actual pursuit of social science. Although my claims are quite new, my strategy turns out to be a traditional one among philosophers of social science.

I will not in this work devote much attention to expounding or defending any particular version of empiricism and physicalism, but shall simply assume that some particular version, pershaps one not yet even expounded or defended, is correct. The general lines of these two "isms" are clear enough, and the precise and defensible expression of these views is clearly among the most important tasks in philosophy. The failure of so many attempts to coherently expound these philosophical theories, let alone defend them, in the face of counterexamples and counterarguments reflects the enormous difficulty of such exposition and defense. But for present purposes such an exposition and defense need not detain us. For, whatever niceties, qualifications, and intricacies characterize the successful enunciation of the "correct" versions of these positions, these versions will still share with all their technically objectionable predecessors and competitors the same implications for the standards against which social science is to be assessed. Therefore, in detecting and deploying these standards, we need not choose as definitive any one of the available alternative formulations of empiricism or physicalism. Nor need we commit ourselves to the view that such a version of each of these theories is likely soon to be available, just so long as we can be sure that somewhere and sometime a canonical version of these theses will be available. In answer to the question, How can I be sure of this eventual availability? I can only answer with autobiography: the brilliant feats of prediction and control embodied in Western technology have caused me to believe that the science which sustains it constitutes knowledge, and that nothing counts as knowledge unless it too can sustain comparable practical results. Furthermore, I believe that the only possible explanation of science's power to sustain such achievements must be empiricist and physicalist in character. I am incapable of seeing how we could have "spun" our science out from a sort of rationalist reflection on the mind's own powers, or through some nonsensory access to the facts of nature. The social sciences would be of only passing interest, only entertaining diversions, like an interesting novel or an exciting film, unless they too stood the chance of leading to the kind of technological achievements characteristic of natural science. For a social science conceived as anything less practical in ultimate application would simply not count as knowledge, on my view. And if it does not count as *knowledge*, disputes about its methods and concepts are no more important than learned literary criticism or film reviews are to our uninformed enjoyment of the books and movies we like.

3

Common Assumptions

If it were said that without such bones and sinews and all the rest of them I should not be able to do what I think is right, it would be true. But to say that it is because of them that I do what I am doing, and not through choice of what is best—although my actions are controlled by mind— would be a very lax and inaccurate form of expression. Fancy being unable to distinguish the cause of a thing and the condition without which it could not be a cause!—Plato, *Phaedo* 99a-b.

One assumption shared by empiricists and their opponents, and indeed by almost everyone who has offered explanations of human behavior, is that distinctively human behavior is to be explained by appeal to various sorts of combinations of the joint operation of beliefs and desires. That is, aside from species of behavior that we share in common with other organisms—like reflexes, for instance—our behavior is purposive and is to be explained by the citation of reasons for it, which reflect choices governed by our purposes and our beliefs about the means to attain them. This is an assumption as ancient as any in the catalogue of received opinions. Indeed, it was already old when Plato insisted, in the *Phaedo*, that human action must be thus explained, and cannot be accounted for merely by the elaboration of immediately prior and concurrent physical states of the agent's body. In fact, Plato offered this view in connection with his critique of the sort of explanations of natural phenomena that Anaxagoras provided. These explanations proceed largely by the enumeration of mechanical or efficient causes, and Plato stigmatizes them as unsatisfactory because they fail to show what good or purpose is served by the occurrence of the phenomena allegedly explained. In this portion of the *Phaedo* at any rate, Plato seems to embrace the view that the only satisfactory explanation of anything whatever is a teleological one. The history of Western science, however, seems to have more fully substantiated Anaxagoras than Plato, for this history may usefully be viewed as the progressive restriction of the domain of irreducibly teleological explanation from the universal scope that Plato accorded it to the narrow scope of the social sciences alone. It is a commonplace to account for primitive explanations of natural phenomena by pointing out that they reflect the phenomena's assimilation to the patterns of explanation by purpose and reason, patterns that are historically and geographically ubiquitous in the accounts of human activities. In the course of the development of Western science, however, Aristotle's teleological physics gave way to Newton's mechanistic, nonpurposive explanations of motion, and eventually of all those phenomena that are ultimately explicable in terms of motion. Thus, after the eighteenth century the domain of teleological explanation was restricted to phenomena now treated in biology and the life sciences. But the succession of advances from Darwin's exposition of the theory of natural selection and Mendel's account of heredity to their synthesis and the emergence of a chemical theory of genetics has eliminated even in these fields any need for distinctively teleological explanation. That is, although purposive or functional explanations are still to be found in biology, these accounts are not treated as autonomous from and fundamentally irreducible to explanations of a mechanistic kind. Although the details of these reductions are controversial in the philosophy of science (and indeed the difficulty of completely spelling them out will be an important feature of the argument of this work), the sketches of this reduction already available, coupled with the empiricist's wider philosophical commitments, make him confident that this characterization of the progressive elimination of irreducible teleological explanations from natural science is correct. In fact, the empiricist is inclined to go further and claim that to the extent an attribution of functionality or purposiveness to an item cannot be explained away at least in principle, that attribution is suspect at best and illegitimate at worst. This is why so important a part of the empiricist's program has involved attempts to provide a nonpurposive foundation for the central case of purposive behavior: human agency. While distinctively teleological forces have been expunged from wider and wider regions of physical and biological science, they remain a fixture in the explanatory network of the social sciences. And this is no surprise, for not only are concepts like desire, aim, want, purpose, intention, hope, expectation, belief, and fear so pervasive and so immanent that they provide the original, primitive scheme of concepts for the explanation of all phenomena, social and natural, but they are also so woven into the moral, religious, legal, and aesthetic fabric of human life that no one can long rid himself of the habit of employing them to describe and explain his own and other's behavior. In this respect Plato exhibits, in the *Phaedo*, a fundamental conviction which almost all sides to every dispute in the history of social science have embraced: the conviction that it is ultimately to such purposive concepts that these subjects must appeal. This is why Mill, for example, devoted the opening passages of his treatment of the philosophy of the social sciences to the issues of free will and determinism, "Liberty and Necessity." For he was committed to treating reasons as *the* determinants of action and causal determinism as the only allowable mode of appearance and operation for these determinants. Accordingly, he needed either to undercut or to blunt the criticism of his view that it was incompatible with the sort of freedom in our choices that is introspectively obvious and basic to attributions of responsibility.

Now it may be supposed that there are at least some important figures in the history of social science who have not embraced this assumption that the explanation of human behavior is to be found in its causal determination by desires and beliefs. Indeed, it might be argued that at least one towering figure of contemporary sociological theory—Emile Durkheim—explicitly disavowed any appeal to the conventional circle of explanatory notions. Moreover, the tradition of sociological and anthropological theory that derives from his findings and his methodological prescriptions purports to have superseded this very conviction about the causal determinants of behavior. In order to show how pervasive this common assumption really is, and how difficult it is to consistently abjure it, the present chapter is devoted largely to showing that despite their disclaimers, Durkheim and his avowed followers ultimately embrace implicitly the very assumption which they reject explicitly. Our examination will reveal the empirical and conceptual lengths to which a social scientist is driven in his effort to avoid the presumption that human behavior is determined by the psychological states common sense and ordinary language identify and appeal to. Additionally, it will provide an opportunity to examine at least some of the arguments and some of the motivation for the claims of methodological holism for which Durkheim and his successors are so well known.

Durkheim undertook his study *Suicide* in part to substantiate a number of highly controversial claims about the subject matter and the methods of sociology. He is of course famous for arguing that sociology is a subject autonomous from and not reducible to psychology on the grounds that it treats "facts" which "must be studied as things, that is, as realities external to the individual." These "facts," institutions, rules, social structures, mores, and so on, are "real, living, active forces which, because of the way they determine the individual, prove their independence of him; which if the individual enters as an element in the combination whence these forces ensue, at least control him once they are formed."[1] The lasting influence of Durkheim, however, is not so much the result of his conceptual arguments for these claims as it is the fertile work in sociology which they accompany and which provides the really compelling grounds for them. In brief, Durkheim offers statistical evidence to show that the incidence of suicides among geographically, occupationally, religiously, educationally, domestically, and economically homogeneous groups varies in systematic ways which demand explanation. Durkheim wants an explanation of why the rate of suicide should differ significantly and synchronically as between Scandinavian and Mediterranean peoples, Catholics and Protestants, married and single persons, elite troops and conscripts; and why it should vary diachronically in unexpected ways: why it should rise during periods of economic and political change, no matter whether the change is for the "better" or not, and fall during periods of stability. The presumptive cause of individual suicide and, by aggregation, of the rate of suicide is in fact recorded with suicide statistics "under the title: *presumptive motives of suicide*" and as Durkheim notes, "it seems natural to profit by this already accomplished work and begin our study by a comparison of such records. They apparently show us the immediate antecedents of different suicides; and is it not good methodology for understanding the phenomenon we are studying to seek first its nearest causes, and then retrace our steps in the series of phenomena if it appears needful?" But after noting the general problem of determining human volitions, especially after the fact of suicide, Durkheim goes on to claim that

> even if more credible, such data could not be very useful, for the motives thus attributed to the suicides . . . are *not their true causes*. The proof is that the proportional number of cases assigned by statistics to each of the presumed causes [poverty, family troubles, debauchery, physical pain, love, jealousy] remain almost identically the same, whereas the absolute figures, on the contrary, show the greatest variation. In France, from 1856 to 1878, suicide rises about 40 per cent, and more than 100 per cent in Saxony in the period 1854-1880. . . . Now in both countries each category of motives retains the same respective importance from one period to another. . . . But for the contributory share of each presumed reason to remain proportionately the same while suicide has doubled its extent, each must be supposed to have doubled its effect. It cannot be by coincidence that all at the same time become doubly fatal. The conclusion is forced that they all depend on a more general state, which all . . . reflect. This it is . . . which is thus the truly determining cause. . . . We must investigate this state without wasting time on its distant repercussions in the consciousness of individuals.[2]

Thus, Durkheim concludes, "we shall try to determine the productive causes of suicide directly, without concerning ourselves with the forms they can assume in particular individuals. Disregarding the individual as such, his motives and his ideas, we shall seek directly the states of various social environments (religious confessions, family, political society, occupational groups, etc.) in terms of which the variation of suicide occur. Only then returning to the individual, shall we study how these general causes become individualized, so as to produce the homicidal results involved."[3] It is important to recognize that Durkheim offers a factual *argument* to show that we should ignore ordinary motivational factors—reasons, purposes, or desires—in searching for an explanation of suicides. He does not simply begin with the tendentious assumption that recourse to such factors is excluded. Indeed, he begins, along with the rest of us, by assuming that such factors *will* provide an explanation. And this shows to be wrong the criticism of Durkheim, proffered by Peter Winch and Alasdair MacIntyre among others, that "Durkheim is forced by his initial semantic decision to the conclusion that the agent's reasons [for suicide] are . . . never causally effective."[4] It is quite true that Durkheim does begin by defining suicide for the purposes of his study in a way that leaves it an open question whether the causes of suicide are the joint operation of belief and desire, but this is only natural in the light of his aims: to explain the incidence of suicide and its relation to other demographic and sociological variables. Durkheim's definition of suicide runs as follows: "Suicide is applied to all those cases resulting directly or indirectly from a positive or negative act of the victim himself, which he knows will produce this result."[5] This definition, it has been suggested, *a priori* preempts the citation of reasons as the cause of suicide. Thus, MacIntyre writes that Durkheim, propounding this definition,

> ignores the distinction between *doing X intending that Y shall result* and *doing X knowing that Y will result*. Now clearly if these two are to be assimilated, the roles of deliberation and the relevance of the agent's reasons will disappear from view. For clearly in the former case the character of Y must be central to the reason the agent has for doing X, but in the latter case the agent may well be doing X either in spite of the character of Y, or not caring one way or the other about the character of Y, or again finding the character of Y desirable, but not desirable enough for him for it to constitute a reason or a motive for doing X. Thus the nature of the reasons *must* differ in the two cases, and if the two cases have the same explanation the agent's reasons can scarcely figure in the explanation.[6]

MacIntyre is right to note that Durkheim's definition does not distinguish between cases in which an agent only believes that a result will obtain and those in which he also desires this result, but he is wrong to claim that slurring over this distinction obliterates the roles of deliberation and reasons among the determinants of an act like suicide. For the definition does not commit us one way or another on their causal relevance; it is so phrased as to be neutral on the question of what the causes of suicide are, as well it should be if the definition is offered as the starting place of an open-minded inquiry as to what the causes of suicide really are. MacIntyre is also correct in his conclusion that *if* the two cases do have the same explanation, then the agent's reasons play no role in the explanation. But this conditional claim plainly does not follow from Durkheim's definition (it has independent methodological warrant in the maxim "same cause, same effect"); still less does its consequent follow from Durkheim's definition of suicide. Thus, MacIntyre is quite wrong in the claim quoted above that Durkheim's "initial semantic decision" precludes the appeal to agents' reasons in the explanation of their behavior.

It is worth noting that Durkheim's empirical argument for this conclusion is in fact not very compelling, and shows, not that the reasons cited in coroners' reports for suicides do not constitute their causes, but at best that they do not constitute the causes for the detected changes in the rate of suicide among certain demographically circumscribed groups during different periods of the nineteenth century. Moreover, this argument is blatantly incompatible with the very next statistical argument that Durkheim employs to establish the same conclusion. "No two occupations," Durkheim claims,

> are more different from each other than agriculture and the liberal professions. The life of an artist, a scholar, a lawyer, an officer, a judge has no resemblance whatever to that of a farmer. It is practically certain then that the social causes for suicide are not the same for both. Now not only are the suicides of these two categories of persons attributed to the same reasons, but the respective importance of these different reasons is supposed to be almost exactly the same in both. . . . Thus, through consideration of motives only, one might think that the causes of suicide . . . are of the same sort in both cases. Yet actually the forces impelling the farm laborer and the cultivated man of the city to suicide are widely different. The reasons ascribed for suicide, therefore, or those to which the suicide himself ascribes his act, are usually only apparent causes.[7]

This argument does proceed by assuming that the causes of suicide differ between the two homogeneous classes in question, and implies that since there seems no difference among the sorts or proportions of different motives for suicide among them, that these motives cannot be causally effective. But aside from begging the question, here the conclusion trades on the denial of the methodological maxim of "same cause, same effect" that its predecessor assumes.

For our purposes, however, the real importance of these arguments is that Durkheim felt impelled to supply factual considerations in favor of his claims about the irrelevance of reasons to the explanation of at least one sociologically important species of human behavior. Those who argue without appeal to empirical data that such a claim is *ipso facto* mistaken and that any account of suicide which proceeds from it has no import for social science are at least implicitly committed to some species of epistemological rationalism. For if, like Winch, they argue that distinctively human actions are those whose determinants must be found among the agent's motives and reasons, without providing evidence of roughly the same kind as Durkheim's, then it is they who are guilty either of question-begging "semantic decisions" or of acquaintance with presumably synthetic a priori truths. For, in the first case, if suicide, or any other distinctively human behavior, is defined as the outcome of deliberation or reasons, then its explanation is preempted in just the way Durkheim's definition is (wrongly) accused of doing. And, in the second case, if we can know, without empirical evidence, that the determinants of actual cases of suicide are of the very sort which Durkheim rejects, then our knowledge of this fact can be founded only in the truth of some version of rationalism. (Unless, of course, our knowledge of this truth is simply a reflection of our definition of suicide as an action—i.e., as a purposive consequence of the operation of beliefs and desires—in which case, the empirical issue is simply pushed back to one about whether there are actual cases of suicides, and the occurrence of the deaths reported in the coroners' reports remains open to explanation, an explanation which Durkheim can still purport to provide.)

I have dwelt on this claim of Durkheim's both because it is controversial and, I think, misunderstood, and because the conclusion to which I shall eventually come, about the bearing of reasons on the explanation of human behavior, is much like Durkheim's, and is also an empirical claim, like Durkheim's. Despite his claims to the contrary, however, Durkheim did in fact implicitly accept the view that social facts or the individual human behavior which they determine are consequences of the operation of motives and reasons. Showing this will reflect how deep is the grip of this explanatory model even among those who think to reject it.

Durkheim, then, would have us treat the motives associated with particular suicides as something akin to epiphenomena, but epiphenomena of what? Durkheim explains the different rates of suicide he detected, and the individual suicides which compose them, in terms of the degree of social integration characteristic of the sociologically homogeneous milieus associated with these different rates. In order to explain why rates of suicide vary in the way they do, either synchronically between these groups or diachronically within one of them at different times, Durkheim hypothesizes three distinct sorts of suicide, each distinguished as an effect of extreme forms of social integration and nonintegration: egoistic suicide, altruistic suicide, and anomic suicide. Presuming a sort of implicit equilibrium model of society in which the social constraints on individual behavior are optimal when they minimize the suicide rate, Durkheim explained his statistical findings by arguing that those demographically or sociologically homogeneous groups with the highest suicide rates were just the ones in which societal constraints on individual conduct were either extremely small, extremely great, or highly uncertain in their character or content. Thus, the relatively larger number of suicides among Protestants (compared with Catholics); among widowed, divorced, single, or childless persons; or in societies not unified by their members' absorption with great changes or momentous events is caused, not by a systematic increase in the desire to end life, but by the weakening of extrapersonal social constraints. This is what Durkheim called egoistic suicide. Altruistic suicide is reflected in the heightened suicide statistics for certain occupations that produce social groups so highly integrated that they force suicide on those of their members who fail to fulfill the group norms and standards. Thus, the military shows a higher suicide rate than civilians, and this rate is especially higher among officers than enlisted men, among elite troops rather than conscripts, among longer-serving soldiers than short-term ones. And anomic suicide explains the increase in rates of suicide during periods of great change, either for the better or worse. Thus, suicide rates increase both during periods of depression and, surprisingly, during boom times as well, while they seem to remain stable during periods of economic stability, whether at prosperous or unprosperous levels. The explanation Durkheim gives in these cases is that economic changes break down the harmony between individuals' needs and wants, and their means of satisfying them. The individual's needs and means are fixed by the role which society accords; once habituated to these roles and their associated norms, members whose means suddenly change, for better or worse, find themselves in new roles, new stations in society, for which they are unprepared and the constraints on which are unknown to them. This normlessness, or "anomie," is the cause of increases in the suicide rate during such times.

Notice that these hypotheses about social integration and causally distinct forms of suicide—even if we accept them—explain only differences in the incidence of suicide among different classes of individuals and are heterogeneous in their explanatory force. Egoistic and anomic suicide explain their phenomena in terms of lessening constraints on self-killing, while altruistic suicide explains the phenomenon as forced on its victim by external constraints. The question remains, What is the *mechanism* whereby variations in the degree of social integration determine variations in the rate of suicides for groups of a given sort? Durkheim, in effect, has uncovered a number of generalizations of a macrosociological kind, and these he in turn explains by appeal to more general hypotheses about how suicide rates vary with degrees of social integration. But underlying these explanatory hypotheses there must be a further account of the mechanism which explains the relation of these variables. Compare the situation in thermodynamics. Here we find a number of interconnected macrophysical variables: pressure, temperature, and volume. For a large number of different gases, changes in each of these variables are directly or inversely proportional to changes in the others of them, although the rate of change varies from gas to gas, as does the range of the values of temperature, volume, and pressure within which they co-vary in this neat way. Data reflecting these relations correspond to the statistical tables of suicides per hundred thousand of population for various sociologically distinguishable groups. And corresponding to the general hypothesis that these data reflect a functional relation between the degree of social integration of these groups and the suicide rate is the equation of state for an ideal gas:

$$PV = rT,$$

where P is pressure, v is volume, T is temperature, and r is a constant that varies with the gas under consideration. But the provision of this general formula, which subsumes the data of a large number of different gases, raises more questions than it answers. For one, why do all gases behave in accordance with it across a certain range of values of its variables; for another, why just across that range and not a wider one? Why do only gases obey this law, and not also solids or liquids? More fundamentally, what is the causal mechanism that relates changes in one of these variables to changes in the others? Does just one mechanism relate the six different variations that the formula comports? These latter questions are the most fundamental, for they must be answered in order to provide answers to the former questions as well. The kinetic theory of gases, of course, provides the answer to all these questions by describing a mechanism involving appeal to the behavior of the microphysical constituents of gases: it treats them as molecules behaving in accordance with the laws of Newtonian mechanics, showing how the aggregation of each molecule's Newtonian properties, mass and velocity, is determined by and determines the values of the thermodynamic properties of the gas of which it is composed. Thus the equation of state for an ideal gas, and all the data which it in turn subsumes, is explained by appeal to a further hypothesis about the underlying mechanism and constituents of gases. In fact, the more we learn about these constituents, the further we can extend the range of values of pressure, temperature, and volume of gases open to explanation.

In this case, appeal to underlying mechanisms is not only demanded by explanatory exigencies, but it also provides the best case possible in the history of science for the empiricist's reductionist imperative. Just for that reason, an exponent of Durkheim's view might challenge the appropriateness of any analogy between the equation of state for an ideal gas and Durkheim's regularities about social integration and suicide. After all, the analogy demands that just as explaining the gas law proceeds by way of appeal to the constituents of gases, so explaining the suicide regularities must appeal to the constituents of society—individual agents and the forces, evidently psychological, determining their individual actions. But to do this seems to be tantamount to a rejection of Durkheim's principle that "whenever a social phenomenon is directly explained by a psychological phenomenon we may be sure that the explanation is false."[8] Nevertheless, the connections between rates of suicide and degrees of social integration are plainly not self-evident: Durkheim clearly took them for an important empirical discovery, and it is equally clear that the connection between these two sorts of variables is not causally direct, but must be mediated by some causal chain. To reject the search for such a chain is to forswear science. I have said that the forces which mediate the relation between degrees of social integration and individual suicide must evidently be psychological. This does not mean that they must involve appeal to just that constellation of beliefs and desires of agents collected in coroners' reports, whose accuracy and relevance Durkheim rejects. For all we know, the mediating factors may be of a purely nonconscious sort, operating in the way that, say, lithium imbalances bring on feelings of depression. But although this sort of causal chain is a possibility, it clearly is not the one that Durkheim implicitly embraces. He implicitly embraces the same old sorts of factors that he hoped to make a clean breast of in the explanation of suicide: beliefs and desires. Consider first anomic suicide, the result of the breakdown of the harmony between an individual's needs and his means for satisfying them, either because his means suddenly far exceed what is necessary to meet his wants, or because his wants suddenly exceed the available means. But clearly, the imbalance required to produce suicide in such cases is not between needs and means, but between means which the agent *believes* himself to have, and needs which he recognizes, that is, ends which he *desires*. Neither Durkheim nor anyone else would construe death from a vitamin deficiency as the result of anomic forces just because the agent who needed a given vitamin had not the means to make good his deficiency. Otherwise many deaths due to malnutrition would have to be recorded as the result of social forces, on Durkheim's theory of anomie. It takes no more imagination to describe cases of death which resulted from a surfeit of means to satisfy given ends, but which could not possibly be treated as suicides. Thus, unless we are to understand the discordance between means and needs as really one between desires and beliefs about available ways of fulfilling them, we shall have to treat most fatal drug overdoses and most deaths due to excessive driving speed as anomic deaths; but surely some of these deaths are sociologically accidental. Accordingly, we must conclude that the underlying mechanism in the case of anomic suicide, at any rate, involves the very causal variables that Durkheim eschewed. The case is no better for egoistic suicides. Unlike anomic suicide, these are cases in which insufficient social integration merely lessens the constraints on suicide, instead of forcing it upon its victim. How can varying degrees of social cohesion do this; what mediates their causal force on individual behavior? Differences in the degree of social cohesiveness determine differences in the expectations of individuals about the behavior of others and about the effects of their own behavior on others, differences in the degree of uniformity and coherence of individual desires and wants, and differences in the degree that individuals' desires involve the condition of others. Otherwise, we might have to categorize the death of a widower *before* learning of his loss as at least in part the result of lessening social forces; we should in consistency admit that a man is insulated from egoistic suicide because he has children, even though they are all illegitimate and unknown to him.

The very notion that our behavior is constrained and even determined by the roles we play, by the social institutions that surround us, and by the implicit and explicit sanctions that govern our lives, to which Durkheim attributes an existence independent of us and our psychological states, presupposes the causal force of these states; for the social sanctions cannot function unless they are mediated by these states, unless we are conscious of them or their consequences, and desire to avoid the costs or reap the rewards of obeying them. For the social facts to which Durkheim is committed do not simply reflect brute regularities which we behave in accordance with, but reflect rules that individuals obey, because of their desires and beliefs. The explanation of why their beliefs and desires result in agents' obeying (or breaking) the rules that express social facts must, on the empiricist reductionist's view, ultimately appeal to brute regularities, like those of natural science. But anyone who finds the explanation of one set of macrosocial facts in the citation of another set must eventually commit himself to an underlying mechanism to connect them; and when this mechanism consists in moral constraints and the appropriateness of behavior to circumstances, as it does with Durkheim, the appeal to psychological states becomes almost unavoidable.

If we turn to Durkheim's intellectual descendants, we can plumb the depths to which avoiding this conclusion must take someone. Among these successors to Durkheim, currently the most fashionable are the structural anthropologists, who take their most direct inspiration from the works of Lévi-Strauss. And indeed, Lévi-Strauss may himself be instanced by some as another important figure in social science who has not embraced the common assumption that the determinants of behavior are to be found in desire and belief. This claim finds its most clear-cut exposition in a controversy that so plainly highlights the current issue that one is inclined to think that if it had not already existed, I would have been forced to invent it. In 1955, George Homans and David Schneider published *Marriage, Authority and Final Causes*,[9] in which they purported to provide what they called "an efficient cause theory" to explain anthropological data which Lévi-Strauss had explained by appeal to what they called "a final cause" theory. The monograph was widely acclaimed, but soon enough elicited a stinging rejoinder by one of Lévi-Strauss's exponents, who attempted to show that in Homans and Schneider's work, the "conclusions are fallacious, its method unsound, and the argument literally preposterous." This rejoinder appeared in 1962 and was aptly titled by its author, Rodney Needham, *Structure and Sentiment*[10]—aptly titled because the real controversy here is between social scientists who accept the common assumption that beliefs and sentiments determine human behavior, as opposed to those who reject it in favor of something they call social structure and declare autonomous from sentiment. As we shall see, proponents of this latter view must in fact also ultimately appeal to "sentiments" and other psychological considerations in order to expound their theory, despite their conviction, cited from Durkheim with a flourish as the last sentence of Needham's work, that "whenever a social phenomenon is directly explained by a psychological phenomenon, we may be sure that the explanation is false."

Kinship, anthropologists are eager to allege, is the most fundamental feature of human society, and kinship algebras, axiomatic presentations of the general form of kinship systems that various geographically and temporally separated peoples share, seem to be the chief ornament of what little theory anthropologists are willing to admit to adopting. It is certainly a matter of considerable interest that primitive societies with no possibility of communication, either because of geographical barriers, or because they are separated in time as well as place, should employ marriage rules reflecting kinship systems that, at some appropriate level of (nontrivial) generality, are roughly, and sometimes almost exactly, the same. More specifically, we can find cases of different tribes with no connections to one another (because, say, one is Australian and the other Asian) whose systematized marital prescriptions and proscriptions are but differing interpretations of the same abstract calculus. This is a general fact that cries out for explanation. And it is about how to explain appropriately the details of this fact that the dispute emerges between reductionists like Homans and Schneider, and structuralists like Lévi-Strauss and his protégé Needham. Of course, the dispute about what general strategy is to be followed in the explanation of kinship structures is rightly pursued only in much narrower and more specialized terms, for otherwise it threatens to degenerate immediately, instead of only eventually, into differences of an epistemological, and therefore unbridgeable, nature. In this case, the dispute devolves on the question of how we are to explain why it is that among small sets of societies in which unilateral cross-cousin marriage is the rule, matrilateral far predominates over patrilateral cross-cousin marriage. Now this is a question so narrow and technical that its very meaning must be explained to all but those who follow the intricacies of structural anthropology, and thus it bids fair to be a factual and not a conceptual question, one properly within the ambit of social science, and not philosophy.

Matrilateral cross-cousin marriage exists when members of a group prefer or expect that males will marry cousins on their mother's side of the family, e.g., mother's brother's daughter, and disapprove marriage with cousins on the father's side, e.g., father's sister's daughter. Patrilateral cross-cousin marriage is the opposite scheme of expectations and disapprovals. The explanandum phenomenon here is why among those societies evincing cross-cousin marriage preference or prescription, matrilateral cross-cousin marriage seems four times as frequent as patrilateral cross-cousin marriage. Lévi-Strauss's explanation, quoted by both parties to the dispute over its acceptability, proceeds as follows: "If, then, in the final analysis, marriage with the father's sister's daughter is less frequent than that with the mother's brother's daughter, it is because the second not only permits but favors a better integration of the group, while the first never succeeds in creating anything but a precarious edifice made up of juxtaposed materials, subject to no general plan, and its discrete texture is exposed to the same fragility as that of each of the little local structures of which it is ultimately composed."[11] Matrilateral marriage leads to greater integration because it results in the transfer of women of one generation from one lineage to another, which then must transfer its women in the next generation to still a third, until the cycle is completed by the remaining lineage, n, in a group transferring its women of generation n to the first lineage; on the other hand, among societies where patrilateral marriage predominates, transfers are reciprocal in that a lineage transfers women to the lineage group which provided its women in the previous generation. Lévi-Strauss calls these transfers exchanges (even though matrilateral transfer is not reciprocal exchange, and in that lies its strength). This explanation Homans and Schneider describe as a "functional" one, laying stress throughout their argument on Levi-Strauss's use of the expression "better social integration"; and they suggest (with textual citations that Needham disputes) that, according to Lévi-Strauss, the underlying mechanism whereby this more beneficial social arrangement came to predominate was not natural selection, or the expression of direct and immediate biological need, but the intelligent recognition among members of these societies that matrilateral marriage is better for their groups because it fosters social integration.

Homans and Schneider are correct to insist on the provision of an underlying mechanism in order to substantiate a functional explanation, and Needham too accepts its necessity. He, however, finds the underlying mechanism to which Lévi-Strauss is committed in an amalgam of selective forces operating on "the unconscious production of [quoting from Lévi-Strauss] 'certain fundamental structures of the human mind' by virtue of which human communities tend . . . to 'integrate and disintegrate along rigid mathematical lines.' "[12] But no matter what the underlying mechanism required by Lévi-Strauss's functionalist phraseology, Homans and Schneider propose to show that no functional explanation of the predominance of matrilateral marriage over patrilateral is required because this predominance can be explained simply by appeal to the ordinary conscious psychological states of members of the groups which practice these prescriptions. Of course, the explanation of the greater frequency of matrilateral cross-cousin marriage they attribute to Lévi-Strauss also involves these same sorts of variables, but their variables' connection with kinship behavior is mediated, on their interpretation of Lévi-Strauss, by the marriage rules, whose functional desirability is recognized by those who obey them. In contrast, they suppose that no appeal to these rules need be made in order to explain the behavior in question, so that the rules themselves do not constitute an "underlying structure" determining the behavior, but simply express regularities in the behavior and are but descriptions of the explanandum in question.

Regardless of whether Homans and Schneider's ascription to Lévi-Strauss is in order, no serious structuralist will accept the attributed view as an acceptable explanation of the phenomena in question. For the principal theoretical motivation of structuralism is the hope that it will provide a method independent of appeal to time-worn variables like conscious beliefs and desires in the explanation of human behavior. Its foundations in the linguistic theory of Ferdinand de Saussure, cited as the touchstone of all subsequent structuralist theory, involve the attempt to determine the causal units of verbal behavior not in semantic content, philological roots, and etymology, nor in the physical character of speech, but in sounds and symbols that are behaviorally homogeneous—are involved in the same causal relations of speech production and recognition—so that their identities are determined by the general regularities in which they figure. These regularities, according to Saussure, and not the semantic content of the units of speech, provide a structure of linguistic *units* independent of individual linguistic peculiarities and the recorded history of language which governs their appearance, their juxtapositions, and the changes in their linguistic roles. Taking inspiration from this account of the fundamental units—the basic subject matter—of linguistics, the structuralist in other social sciences hopes to find the basic units, the theoretically significant variables of his subject not in the declarations of his subjects, nor even in items open to description in the language of his subjects, but in the objects that figure in regularities about his subjects which he may discover independent of their own accounts of the matter. Thus, in structural anthropology the kinship units turn out not to be the ones which are encapsulated in our own language (like mother, uncle, niece, etc.), nor even the ones that figure as basic in the language of a society with a well-articulated set of marriage prescriptions and proscriptions; the real units of kinship are those which figure in the general rules of kinship systems that the anthropologist detects to be operating within and across diverse societies. In effect, the structuralist explicitly surrenders the common assumption that the determinants of the human behavior that he studies are the ones which common sense dictates. He may well not yet know what they are, but his method enables him to categorize the phenomena he studies in a way that cuts across the categorizations of ordinary language, and thus potentially frees him from dividing up behavior into intentional action as opposed to unintentional movement. Thus, if the anthropologist discovers the same abstract set of kinship rules operating in different groups, with different rationales, or with no articulated rationale or even recognition, he is in a position to say that he has uncovered a form of behavior not open to explanation in terms of conventional human purposes and beliefs, but requiring description and explanation in terms that transcend common assumptions. Such an attainment represents the ideal, at any rate, on which the structuralist who follows Saussure's example keeps his eye.

Homans and Schneider, however, propose to show that such flights into structuralist theory are superfluous, for we can explain the phenomenon without them. Their explanation for the higher proportion of matrilateral over patrilateral cross-cousin marriages turns on the assumption that persons want to marry the children of others for whom they have positive feelings, combined with an anthropological fact about lineages. Among lineages, lines of descent, we can distinguish patrilineal and matrilineal variants, that is, systems which trace descent through the father's side of a family as opposed to systems which do so through the mother's side of a family. Homans and Schneider contend that the linearity of a line of descent determines what they call the "locus of jural authority." That is, when descent is traced through the mother's side of a family, established authority over younger members of the society is typically found in the mother's family—her brother, or uncle, etc. And where societies are patrilinear, the locus of jural authority, the focus of respectful distance, obligation, discipline, sanction, etc., is to be found on the father's side—indeed, is typically the father himself. By contrast, if one side of a family represents authority, the other finds itself the locus of indulgence, affection, friendship, fondness, and easy company. It is in agents' beliefs about these opposed loci, and in their preferences between them, that Homans and Schneider find the purely psychological mechanism that explains the prevalence of matrilateral cross-cousin marriage. In a patrilineal society, an agent is inclined to look for a wife among available women on his mother's side, among his maternal uncles' children. For this is the side of the family with whom his relations are most intimate and sympathetic. And similarly, in such a society, men will look to their maternal nephews for sons-in-law, for these are the potential grooms to whom they are sentimentally the closest. Thus, in patrilinear societies individuals are inclined to undertake matrilateral rather than patrilateral cross-cousin marriage. But how does this inclination explain the prescription, the norm to so marry, when it exists? The explanation is disarmingly simple: "We hold . . . that norms are not independent of actual behavior. When the social structures of many kinship groups in a society are similar, which is eminently the case with primitive societies, then many individuals will tend to develop similar sentiments and behavior towards similar kinsmen. For example many egos will develop similar sentiments and behavior towards their respective mother's brothers. In time such sentiments and behavior will become recognized as the right and proper ones: they will be enshrined as norms."[13]

Having shown that patrilineages induce preferences for maternal cross-cousins, and matrilineages for paternal cross-cousins, and having suggested that such preferences are what generate norms, Homans and Schneider hypothesize that matrilateral marriage is more widespread than patrilateral because patrilineages are more common than matrilineages. They then test this hypothesis against available anthropological data. Out of a total of thirty-three societies reported in the anthropological literature as practicing unilateral cross-cousin marriage, Homans and Schneider's claim is confirmed by twenty-seven of them. That is, twenty-two of the total are patrilineal-matrilateral; five are matrilineal-patrilateral. Unsatisfied with this degree of confirmation, they go on to show that the apparent counterexamples can be accounted for by appeal to the more general hypothesis from which the lineage-laterality hypothesis follows. They show that all but one of the exceptions are societies in which the locus of jural authority is not vested with the linearity of the society, but is otherwise distributed, and that in each of these cases the laterality of marriage is away from the locus of authority. In other words, both the exceptions and the confirming data are explained on the general claim about preferences of agents for marriage with agreeable and sympathetic kinsmen, rather than with members of authoritarian branches. Homans and Schneider conclude that since the frequency of matrilateral over patrilateral cross-cousin marriage can be explained by appeal to the personal predilections of members of societies practicing these forms of kinship, Lévi-Strauss's explanation of this frequency in terms of "the harmonic character of generalized exchange" and its greater contribution to "organic solidarity" is superfluous, and represents theoretical excess. If they are correct, Lévi-Strauss's attempt to provide an abstract and general theory of human behavior which transcends our commonsense suppositions about its nature and mechanism is at least in this case seriously undermined.

Of course, Homans and Schneider intend to treat this dispute as a test case of the merits of structuralist thinking in social science *in toto*; and from the vigor of Needham's defense of Lévi-Strauss, his method, and its subject's autonomy from commonsense psychological explanation, we may infer that for him too the issues broached transcend the narrow question of why matrilateral cross-cousin marriage is more frequent than patrilateral unions. Needham tells us that "in not one case or respect that I have been able to discover has psychology afforded in itself a satisfying or acceptable answer to a sociological problem. It is from this pragmatic basis that I have set out to expose the fallaciousness of the psychological explanation . . . proposed by Homans and Schneider" (p. 126). Much of Needham's monograph is taken up with factual matters which he claims have been missed or misrepresented by Homans and Schneider, and which undercut the confirmation of their theory. A good deal is also devoted to disputing their textual exegesis of and the emphases they place on passages from Lévi-Strauss. But for our purposes what is more important than these points of disagreement are the methodological and epistemological conclusions to which Needham's argument implicitly and sometimes self-consciously commits him, and his own ultimate commitment willy-nilly to the efficacy of psychological forces in the explanation of anthropological phenomena.

After offering an exposition of the issue between Lévi-Strauss and Homans and Schneider, Needham suggests that for several reasons they are at cross-purposes, because among other things, Lévi-Strauss's "theory is not a causal theory at all." He then turns to disputing some of the anthropological authorities, like A. R. Radcliff-Brown, whom his opponents cite. It is not until Needham's monograph is well under way that he comes to what he calls a "general" argument. It is, as might be expected, one which cites Durkheim's dictum that "the psychological factor is too general to predetermine the course of social phenomena. Since it does not imply one social form rather than another, it cannot explain any."[14] In particular, "if the sentiments in question are so completely general in lineal descent systems, they cannot explain (account for the "adoption," the existence of) a marital institution which appears with such rarity among them" (Needham, p. 50). But aside from the fact that in the explanandum phenomena, the marital institution in question (unilaterality of cross-cousin marriage) so far from being rare is *universal* (all thirty-three societies evince it), the standards which Needham and Durkheim hold up for adequate psychological explanation are impossibly high for any explanatory theory, psychological or otherwise. Durkheim's employment of the expression "imply" suggests that to be adequate, the considerations cited in the explanans must be the sufficient conditions of the explanandum phenomena; otherwise they have no explanatory force whatever. And this is plainly too strong a requirement. Needham's insistence that the more general consideration cannot explain the less general is, without further explanation, equally unsupportable. For the characteristic strategy of scientific explanation is that it proceeds in just this way. Of course, Needham's real complaint is that many societies show lineality, and in all of them, the psychological generalizations about avoiding the children of jural authority is supposed to obtain; consequently, these two facts do not jointly explain why it is that among unilateralized societies, one form predominates; for these alleged facts are compatible with the evident nonexistence of unilaterality in all but the small minority of societies that constitute the explanandum class. This is another version of Durkheim's demand that to have any explanatory force, the explanans must imply, be the sufficient conditions of, the explanandum. Naturally, the closer an explanation comes to providing the sufficient conditions of its explanandum, the more powerful it is, but recognizing this fact is compatible with according some explanatory power to theories which fail to satisfy this ideal completely. Moreoever, insofar as causal explanations never provide the causally sufficient conditions of their explananda (to do so they would have to be indefinitely long), the demand that Homans and Schneider's causal theory meet this condition is unwarranted in any case. Of course it may well be that they have too generally and incompletely identified the sentiments to which they appeal, or have even appealed to the wrong factors altogether, but to reject their theory on considerations of this order reflects differences that can hardly be disputed within the confines of anthropology, for they turn on the nature of causation and the grounds of our causal knowledge.

Needham makes another charge against Homans and Schneider, one more expected from an anthropologist and more familiar to a philosopher, but one which is quite surprising for a defender of structuralist theory: he condemns his opponents' disregard of the system of categories which their theory is intended to elucidate. "This doctrinaire neglect of the facts of the case—here the connotations of the categories for the people who employ them and who order their lives by them—is the gravest defect of their enterprise" (p. 45). Now unless Needham means something very special by the "connotations of the kinship categories," and by their "role in the natives' ordering of their own lives," his criticism of Homans and Schneider is either or both patently groundless and clearly self-contradictory. After all, to talk about the subjective connotations of linguistic categories and about their role in the formulation and guidance of behavior is to admit the causal force of intentions, purposes, meanings, etc., as determinants of behavior. It is, in short, to appeal to the general psychological variables that Homans and Schneider cite and Needham rejects. By the appeal to such categories, Needham must have something different in mind. But it is only when we reach his own positive view about how the explananda in question are to be accounted for, that we begin to see what he has in mind.

Needham begins this account with the truism that "you cannot . . . explain what you do not first understand." But, he goes on more substantively: "The way to acquire such understanding is by a total structural analysis of all the recorded facts on one society. This means surveying every facet of social life, not merely the allocation of authority within the narrow circle of domestic relatives, or even the conventional institutions of 'kinship' and marriage. Such an examination involves a systematic comprehension of the life of a society in terms of the system of the classification employed by the people themselves, and an analysis in terms of relations of the widest generality" (p. 73). Such a study, which Needham pursues in miniature for the Purum people of India and Burma, shows that "marriage with the patrilateral cross-cousin would not be simply an isolated act of wrongdoing having limited (punitive) consequences just for the two individuals by whom it was perpetrated; but it would constitute an onslaught on the entire complex of identically ordered relationships which are the society, and on the symbolic classification which is its ideological life" (p. 99). All of these relations and all of the symbols through which Purum society classifies itself reflect the *generalized exchange* that Lévi-Strauss detects in matrilateral cross-cousin marriage, and to which he attributes comparatively greater "organic solidarity" than other sorts of exchange (such as that involved in patrilateral kinship systems). All of the characteristic behavior and traditions of this society are claimed to reflect the generalized exchange that binds members of a society to one another. The topology of kinship structures is mirrored in ceremonies, in hierarchies of respect and authority, in work, in myth, art, domestic architecture, in all aspects of life. But reflection on the notion of exchange to which Needham and Lévi-Strauss appeal reveals that beneath it lurks that same commitment to the explanatory efficacy of psychological variables that on the latter's behalf, the former is at pains to deny. For "exchange" in this context is not to be understood as the purely mechanical phenomenon of exchange of momentum, for example, familiar in Newtonian physics. Rather, it is to be understood in terms of "the notion of reciprocity," as Needham says, "reciprocity, considered as the most immediate form in which the opposition between self and others can be integrated" and involving a gift whose "transfer from one individual to another changes them into partners and adds a new quality to the valuable which is transferred." As Needham admits, "Stated in this fashion the points are inevitably obscure, but [citing Lévi-Strauss[15]] the central feature is the apprehension by the human mind of reciprocal relations" (pp. 27-28). In effect, Needham is really committed as much to a psychological mechanism underlying and explaining "generalized exchange" as Homans and Schneider are to one underlying unilateral marriage. And if anything, the psychological mechanism to which Needham is committed is more general and more incompletely specified than he accuses theirs of being. Instead of finding one specific set of fairly commonsensical sentiments underlying one particular aspect of social behavior, Needham locates a much more general and more speculative psychological mechanism—the recognition of reciprocity of relations, and the assumed purpose of acting on this recognition—as the explanation of all of the characteristics of a social structure. This commitment, willy-nilly, is clearly revealed in Needham's final chapter when he turns to providing a detailed structuralist explanation for the rarity of patrilateral kinship systems, which derives directly, he claims, from taking Lévi-Strauss's notion of organic solidarity seriously.

Recall that in patrilateral alliance, each group functions "as a group alternatively as wife-givers to another group and as wife-takers from that group."

Now consider . . . what happens when two such groups meet as groups, at an event such as the contraction of marriage in which members of two groups and of two successive generations participate at the same time. It is easy to stipulate in a matrilateral system that B, as wife takers, will acknowledge their inferior status by going to A with masculine presents which express their inferiority, for this is a continuing status occupied by successive generations for as long as the relationship between the two groups lasts. But whatever will happen where this status alternates by generations? Senior members will stand in the relation $A > B$, while the next generation will be in that of $A < B$. If A_1 acts on behalf of his son, A_2, he will be putting himself in a position of inferiority to B_1, his own wife taker, to whom he himself is superior within his own generation; and B_1, acting as custodian of his daughter, will be in the position of wife giver to the son of A_1, his own wife giver. In one status, B_1 should go to his superior, A_1, with masculine goods; in another, B_1 should himself assert superiority by receiving A_1 and accepting such goods from him. And, once again, . . . the relations between these structural positions involve not only the cession of women, but all the items of the system of prestations which comprise the economic and ritual intercourse between units of the society. This, it will be agreed, is a picture of sheer confusion.

. . . a hypothetical patrilateral system would be so obviously difficult to work with as to be reckoned for practical purposes socially impossible. In other words, as a solidary arrangement it would be, to say the least, less effective than a matrilateral system—which is precisely Lévi-Strauss' essential point (pp. 113-15).

Clearly, Needham is excluding the possibility of patrilateral systems on the ground that they are impossible systems for natives to consciously work with, that they represent rules that are simply too confusing to be carried out by members of societies that they might govern. They involve shifts of recognized status and consequent changes in appropriate action too complex to be practically applied. The positions of inferiority and superiority described reflect the beliefs of agents, and the exchanges reflect intentional actions undertaken on these beliefs about relative social position. It is because of the variation of these beliefs and consequent actions required of individual agents from occasion to occasion that, Needham concludes, the patrilateral system is unworkable and ineffective in the provision of social solidarity. Consequently, it is hard to see how he has avoided the recourse to psychology that, quoting Durkheim, he rejects as never providing the explanation of a social fact. Of course, Needham's appeal is to a different range of psychological states than Homans and Schneider's, but it is a recourse to psychology for all that.

Although Needham's ostensible aim is to undercut one particular attempt to explain a social phenomenon by finding a psychological mechanism to generate and maintain it, his real aim is to underwrite the existence and autonomy of a science, a discipline, distinct from and unreducible to psychology. This, of course, was the implicit aim of much of Durkheim's empirical research and the explicit conclusion of his methodological writings. It is a characteristic especially of anthropology and sociology, that its students betray anxiety about the existence of a distinctive subject matter for their subject by arguing so forcefully, and reacting so defensively, when faced with claims to the contrary. This concern with the very existence of their subject and the insistence on its hermetic insulation from the findings and theories of any other subject is particularly taxing for the anthropologist. For he insists also, as Needham exemplifies, on the importance of understanding the behavior of his subjects, of attending to and coming to grips with the terms in which the subjects view their own lives and actions. The insistence that to understand a society the anthropologist must "go native," come to think in the terms of his subjects' language, and reason in accordance with their principles and beliefs, makes the anthropologist's task peculiarly open to annexation by the psychologist, whose domain is ostensibly a general account of cognition and behavior throughout the species. To prevent this annexation, the anthropologist may append to his insistence on the importance of his natives' accounts of themselves the additional claim that these accounts do not really explain their behavior, but, along with their behavior, are further symptoms which are ultimately explained by considerations transcending the narrowly psychological.

In the case of a sociologist like Durkheim the recourse to explanatory considerations transcending those of psychology was based on alleged empirical findings, and not on a logically prior commitment to the denial of physicalist reductionism. Durkheim was himself emphatic on the unity of method between the natural and social sciences, and certainly embraced the tenets of empiricism with regard to the nature and justification of knowledge. His departure from traditional empiricist views consists in his holism, in his claim that applying empirical methods to the explanation of social phenomena reveals empirical regularities, and ultimately nomological generalizations of a special sort: ones whose truth commits us to the existence of societal facts and forces which are not mere aggregations of facts about individual agents and the psychological forces that govern their behavior. The only way fundamentally to refute a claim like this is to show that the explanandum phenomenon can be accounted for without appeal to laws that quantify over such societal facts and forces, thus demonstrating their superfluousness. And this is, of course, what Homans and Schneider set out to do. Whether they have succeeded or not, the question of whether this range of special entities exists, and thereby renders autonomous the subject which purports to study them, remains always empirically open. If Homans and Schneider have failed to show that appeal to them can be dispensed with, it is always possible that someone else will demonstrate this to general satisfaction. Accordingly, the autonomy and independence of a discipline like sociology or anthropology remains perpetually contingent on empirical considerations. Such a contingent autonomy is sometimes insufficient assurance for anthropologists, sociologists, and others who find the reduction of social sciences to psychology, and perhaps ultimately to physical science, unacceptable, or even conceptually incoherent. Such thinkers must ultimately find the substantiation of their stronger claim, that a reduction could never be effected, in the rejection of physicalism as a conceptual possibility and empiricism as the exclusive account of cognitive justification. Needham seems himself committed to just such metaphysical and epistemological claims. His argument is not just that Homans and Schneider have in fact failed to reduce the sociological phenomenon in question to a psychological one, but that their entire attempt to do so is radically misconceived. He writes: "It seems fairly certain that causal arguments in terms of sentiments (or any other psychological factors) have not been of pragmatic value in sociological analysis. But it is not only psychological causes which are at issue. The question is whether causal arguments as such are of explanatory value in the analysis of social institutions"(p. 122). Needham's answer to this question is no. It will be recalled that at the outset of his treatment Needham asserts that Lévi-Strauss's "theory is not a causal theory at all." This mysterious claim becomes explicable if we interpret Needham's position to be that psychologism is a pointless strategy in anthropology and sociology, because these subjects are not empirical disciplines at all. For if the sort of knowledge they seek is not causal knowledge, then their aim is obviously not the provision of empirical general laws, nor are their findings necessarily controlled by observation, experiment, and other methods that empiricism requires for the establishment of such laws. "Causal analysis," Needham writes, "is only one method of understanding: it is not a paradigm of all explanation to which sociological analysis must conform" (ibid.). This strategy enables someone in Needham's position to escape the charge that in denying the adequacy of psychological mechanisms to explain causally the operation of sociological systems, he has himself implicitly appealed to such psychological mechanism. For he may conjoin to his claim that the beliefs and desires that he ascribes to agents are but epiphenomenal manifestations of a whole symbol-system that renders intelligible the lives of his subject, the further claim that understanding this system does not constitute causal knowledge at all. And he may conclude that his citation of psychological factors, like the recognition of reciprocity and status, does not constitute an inconsistent appeal to the very range of variables whose explanatory power he rejects. For he is not appealing to causal mechanisms in citing these factors; rather, he is revealing one particularly accessible reflection of the structure of logically, not causally, interconnected symbols, whose grasp sociological knowledge consists in. This sort of argument will enable its proponent to infer that his subject is not just as a matter of current empirical belief autonomous from other allegedly more fundamental disciplines, but that it is conceptually isolated from these other subjects by logically incompatible methodological differences: these other subjects aim to provide a causal account of the phenomena they treat; his subject aims for a treatment that provides a logically different type of understanding. Therefore, its findings and theories can no more be reduced to, absorbed by, or explained in terms of those of another subject than induction, for instance, can be shown to be a species of deduction.

But if the relations between the items cited in the explananda and the explanans of this allegedly autonomous discipline are not causal ones, then what can they be, and how does the citation of the latter provide explanations of the former, and constitute knowledge at all? There seem to be only two possible answers to this question. One of these answers was popular among philosophers who in the sixties produced accounts of the nature of explanation in social sciences under the influence of Wittgenstein. These philosophers brought together the anthropologist's claim that we must at least initially understand the natives' own symbolic, categorial, and linguistic scheme, with the philosopher's claim that the relations between members of symbolic, categorial, and linguistic schemes are not contingent but conceptual ones. They thereby insured the autonomy of the methods and the concepts of social science from the empiricist pretensions of natural science. Thus, Peter Winch, a principal of this school of philosophers of social science, has written: "To give an account of the meaning of a word is to describe how it is used; and to describe how it is used is to describe the social intercourse into which it enters. If social relations between men exist in and through their ideas, then since the relations between ideas are internal [logical] relations, too . . . [they] fall into the same logical category as do relations between ideas."[16] If, in the present case, kinship rules, ceremonial practices, native categories and language, and the natives' own explanation of their behavior all reflect an underlying structure of logically interrelated units, as Needham would have us believe, then, on the assumption that understanding this structure is what understanding the society under investigation consists in, explanations of features of this structure are not causal explanations. They are conceptual clarifications, akin more to proofs in geometry than demonstrations in physics. Thus, the common assumption that psychological factors like belief and desire causally explain human behavior can be escaped by admitting that beliefs and desires are involved in the explanation of human behavior, but not as its *causes*, rather as its *meaning*, as part of the logical framework within which the behavior, qua human action, fits. This seems to be very much the way in which Needham views the theory he defends: "The essential thing to be appreciated is that [Lévi-Strauss] is concerned with structural possibility. . . . He is not concerned with and says nothing about statistical frequencies" (p. 20). It has also the advantage of establishing on logical grounds, and not merely on factual grounds, the irrelevance of empirical finds in psychology to the aims and claims of anthropology and sociology. But even if such an account of these subjects can be made out, treating their theories as wholly conceptual claims without empirical or, in particular, nomological form, the problem of explaining particular human actions has only been postponed. For now the exponent of such a purely conceptual account must provide a method of showing when any particular event instantiates the kind of categories that his theory trades in. And this raises all the empirical questions about the determinants of human action all over again. Compare the situation in geometry. If we defend Euclidean geometry as a body of necessary truths, necessary because they reflect only meanings, definitions, and the consequences of definitions, then it becomes an empirically open question whether the objects of the theory thus interpreted exist or not; whether, for instance, there are in nature Euclidean triangles. For unless there are, the formalism of geometry, true or not, will have no application to the explanation of any particular events, states, or conditions in nature. Similarly, if we treat a structuralist account of human action as a body of conceptual truths, we may defend its autonomy from any allegedly more fundamental causal account of human action. But if we do this, we must provide ways of connecting the terms of this theory with items in the range of phenomena it is to explain, and these connections will raise all the empirical questions about why one particular event, state, or condition is or is not, as a matter of empirical fact, concomitant with another. Unless we provide contingent existence claims for the variables of our body of necessary truths, and means of determining the truth of these claims, our theory, true or false, will have no more application than uninterpreted pure geometry. It will leave its field open, open to the very disciplines from which it and its subject matter are held to be autonomous.

Of course, one way to guarantee the necessary truth of geometry and its applicability is to treat it as a body of a priori synthetic truths. This is an alternative open only to those willing to embrace rationalism in epistemology. Similarly, one way to defend the methodological and conceptual autonomy of a structuralist theory and its applicability to the phenomena which it is meant to explain is by deeming it a body of synthetic truths, known a priori. But again, this can be done only on the assumption that rationalism is correct. This is the second of the two ways in which one might defend the claim that a structuralist theory, or any other sort of account, provides explanations of occurrent states, events, conditions, etc., even though that theory is not a causal one or one open to assessment by empirical methods. The theory thus treated turns out to constitute knowledge, just because empiricist strictures on what constitutes justified belief are rejected. But if treating the components of a structural account of human behavior as a body of conceptual truths turns out to be insufficient to defend the autonomy of anthropological theories and their subjects from the invasion (or, if I am correct, the presupposition) of psychological causation, then the autonomy of disciplines that embrace such accounts ultimately rests on an implicit commitment to rationalism, as well as to holism.

Lest I be misunderstood, I do not hereby accuse either Durkheim or Lévi-Strauss of commitment to rationalism (though the former is committed at least to methodological holism as an empirical claim). My aim is to show that someone who, like Needham, wishes to reject the assumption that human activity is at bottom the causal consequence of individuals' reasons, beliefs, desires, intentions, fears, hopes, expectations, and the rest, in the end paints himself into a rationalist corner. For unless he does so, his own arguments and their employment of these very variables belie his conclusions, and his rhetorical aims overreach his argument. Naturally, rationalism might be correct after all. But for our purposes, simply showing the commitment of a view to this epistemological position is enough to dispose of it. For we have assumed that some version of empiricism is correct and that, accordingly, any view that entails rationalism is mistaken. Thus, insofar as Needham's attempt to escape the common assumption about the determinants of human behavior involves the embrace of rationalism it must be mistaken. More important, and quite independent of whether rationalism or empiricism is the correct view in epistemology, we can now see what radical conclusions are entrained by contemporary attempts to reject the assumption, common since well before Plato's exposition in the *Phaedo*, that the causal explanation of our behavior is to be found in our beliefs and desires.

4

Empiricism about Reasons

A commitment to empiricism, together with the common assumption that desires and beliefs—reasons for short—are the determinants of human behavior, generates a particular research program. Social scientists who explicitly or implicitly embrace these two views are unavoidably committed to the search for general laws relating reasons and actions. Of course, many social scientists never actively engage in the search for such laws, or for regularities of any other kind, and many of the findings of social science seem acceptable even though they cite no laws, nor aim ultimately at the discovery of any. Nevertheless, in an important sense, if empiricism is correct, the significant findings of social science must rest ultimately on nomological regularities, and the provision of such generalizations is an unavoidable duty for these sciences. In this chapter I want to explain why this is true, to examine the actual search for the requisite laws, and to begin to consider the upshot of the failure of social science to provide them.

The common assumption, which in the last chapter was attributed even to those social scientists who purported explicitly to forswear it, is more than just a common assumption of social science. Our legal system, our moral institutions, our domestic relations, our occupations, and our own consciousnesses are shot through with this assumption. We all believe that at least some, and indeed, almost all of our everyday singular statements about the causes of particular actions and the effects of beliefs and desires held by particular agents at particular times are *true*. Any theory whose consequences involved the assertion that all of these claims are false or unwarranted would simply be rejected out of hand. Indeed, some philosophers have argued that such a theory would be inexpressible or inconceivable.[1] The reader can no

49

more treat seriously the suggestion that the causes of my writing this work did not include some reasons or other, than he can treat seriously the suggestion that his reading it is the effect of no intention whatever (including that of just passing the time by reading something of no interest, for instance). The attribution of reasons in the etiology of some, indeed, many actions, represents paradigmatic explanations of these actions, in the sense that everyone accepts that they cite the real determinants of the actions to be explained. What is controversial about these explanatory assertions is not their truth, but their grounds, the considerations by virtue of which they are true and the particular assertions that must be cited in turn to explain the truth of these singular causal assertions. Thus, much history, biography, and the case studies so widespread in sociology, politics, and other social sciences reflect attempts to uncover the particular causes or effects of human actions and their aggregations; and the claims of social scientists to have uncovered true singular causal claims are substantiated by the firm analogies between their findings and the paradigm cases of explanations in ordinary contexts. Our confidence in the truth of their findings is an inductively warrantable inference from the confidence we have in the explanation of our own actions. But the question remains, for these findings as for the established claims of ordinary life, What are the grounds on which they are true, and what, in turn, explains their truth?

Insofar as these claims are causal ones, empiricism commits us to the existence of laws which justify the singular causal judgements, and whose discovery will explain them. But are they causal claims? The singular claim that Jones climbed the ladder because he wanted to recover his hat from the roof and believed that climbing the ladder was an available means to do so must assert a contingent connection between his action and his states of belief and desire. If the connection is not contingent, but is conceptual, or logical, then the citation of his reasons will leave open the question of why this event occurred. Thus if, following some philosophers of social science, we treat the connection of reasons and actions as noncontingent, then obviously the question of why the action which constitutes the event to be explained occurred remains open. For on the assumption of a necessary connection between actions and their reasons, the occurrence of the action logically entails the operation of the reasons. To see this compare the following case. We ask why a particular table has been made in the shape of a triangle and are told that it has been made in the shape of a three-sided plain figure. The proffered explanation is unacceptable because the stipulative connection between triangle and three-sided plain figure renders the assertion incapable of explaining why the table took on the shape it did, rather than that of a rectangle or circle. When we ask why it took on the shape of a three-sided plain figure, we are asking the same question all over again. Similarly, if citing a person's reasons for acting is, in fact, a citation of features logically or semantically connected to his action, then in asking for an explanation of why these reasons obtained, we are in effect asking the same question all over again, why the event to be explained occurred. If to assert that a particular event is a particular action *entails* that there are specific reasons for its occurrence, then the claim that the event occurred because of the specific reasons is just to say that the event was the particular action whose explanation is sought. But this is just to assert the occurrence of the explanandum event, and not to explain it. If the citation of particular reasons really does explain their associated actions, the connection between these reasons and the action must be contingent, and not logically necessary. Consequently, it must be false that the description of an event as a particular action is logically entailed by, or itself logically entails, the reasons for which it occurs.

This false claim must be sharply distinguished from the truth that the description of an event as an action entails that there are some reasons or other for the event's occurrence. But this truth no more entails a logical connection between particular actions and particular reasons than the evident truth that every effect has a cause entails a logical connection between particular causes and their particular effects. Failure to distinguish the issue of whether every effect has a cause from the question of whether every event is an effect led to much confusion in pre-Humean discussions of causation, and the parallel failure to distinguish the question of whether every action has a reason from the question of whether there are actions (whether some of the events in which persons figure are actions) leads to similar confusions. If the existence of some reason or other is a logically necessary condition of the occurrence of an action, it is an easy if mistaken step to the conclusion that the particular reason which is connected to a specific action is logically necessary for it as well. This mistake is especially easy to make if the *terms* we employ to state the reason and to describe the action are the same or are logically connected. This logical truth, that for every action there is a reason, will, however, explain both our commitment to the truth of at least some of our ordinary singular explanatory statements about actions, and explain why we accept the case studies of social science as uncovering truths about particular historical events, their consequences, and their antecedents even in the absence of generalizations about them. For we believe that there are actions, that many of the events in which persons figure are not just occurrences that happen to them, but are actions. This belief is so firm, so closed to challenge, that the very suggestion that there are no actions but only, say, movements, is one that we cannot coherently entertain, if only because entertaining a proposition is itself an intentional event, an action. And this is why we cannot entertain any suggestion to the effect that all our singular statements about the reasons for actions are false. Since we believe there are actions, we are committed to the existence of collateral reasons for them. But inasmuch as we treat the citation of reasons as explanations for their associated actions, it follows that we are committed to the existence of contingent connections between particular reasons and specific actions.

From this conclusion, that the connection between actions and the reasons given for them is contingent, to the search for laws of human action is but a short step for the empiricist. For on his view the only contingent explanatory connections possible are those underwritten by general laws. The empiricist's argument here turns on his analysis of the contingent explanatory connections among natural phenomena. The empiricist, of course, rejects any justification but experience for our claims about the determinants of events, states, and conditions of various sorts. And among those determinants there is no observable, detectable, or inferable mark or sign that distinguishes the sequences in which they participate from purely accidental ones. The only feature that distinguishes nonaccidental contingent connections from accidental ones is the apparent subsumption of the former under general laws. Thus the empiricist finds that causation among natural objects consists in the manifestation of regularities and establishes as a necessary condition for the truth of any singular causal claim the truth of some nomological generalization or other. In fact, because of the vagueness of causal language and the variability of the standards invoked for causal claims, because of the controversies about whether causes are contingently necessary or sufficient for their effects or some combination of these conditions, because of problems surrounding the asymmetry of causal relations, more than one empiricist has proposed to suppress the expression 'causation' in favor of the more general notion of nomological connection. Perhaps most famous of the proponents of this view is Bertrand Russell, who insisted that since every ordinary causal connection is but the reflection of the operation of nomological generalizations, we could dispense with the word 'cause' in favor of appeals to connections between contingent events in virtue of the truth of contingent laws of nature.[2] More reasonable is the suggestion that, in light of the empiricist analysis of causation, we simply employ the term 'cause' as a cover word for all contingent connections that reflect the operation of general laws. It is in this sense that I shall employ the expression, leaving aside the questions about causation alluded to above as irrelevant to the empiricist program in social science. Thus, if reasons are cited and accepted as factors which really explain the occurrence of particular events that are actions, then the beliefs and desires that constitute these reasons must be logically independent of and contingently though nonaccidentally connected to the event they are cited to explain. And the existence of such connections turns on the existence of general laws to subsume jointly explanans and explanandum. Whether these laws will reveal the desires and beliefs to be sufficient conditions or necessary conditions or some combination of them for their effects hinges on the particular form they take; similarly, temporal and spatial relations between states of belief and desire on the one hand and the events that constitute actions on the other will also be a matter of the special details of these laws. But that there are such laws cannot be doubted, on pain of surrendering either empiricism or the existence of actions (as opposed to mere movement), or the explanatory propriety of the citation of reasons in everyday life and in social science.

Notice, however, that though empiricism and the truth of just one causal judgement about reasons and actions commit us to the truth of some general law or other, they do not commit us to any particular general law. Just as the assertion that an event is an effect commits us to the existence of a cause but does not restrict us as to which event is its cause, just as calling a movement an action commits us to the existence of some reason or other for it but not to any particular one, similarly, citing a particular reason as the cause of a particular action does not commit us to any one of the indefinitely many lawlike statements we could construct to subsume them. It only commits us to the truth of one or another member of this set. This explains how it is possible for many an empiricist social scientist to focus on establishing singular causal claims and not trouble himself to search for general laws to underwrite them. The work he undertakes is perfectly consistent with the dictates of empiricism, even if it finds its rationale outside the methodological dictates of empiricism. On the other hand, the joint commitment to empiricism and to the explanatory relevance of reasons explains much of the actual character of those sciences that have self-consciously searched for the general laws required by their commitment. Thus, while all the social sciences presuppose that the events they propose to explain are explained by the provision of reasons, some of the social sciences either explicitly stipulate laws or sustain searches for laws underlying the claims of these other subjects. The clearest example of such a science is economics. Indeed, the history of economic theory is the history of the sustained search for laws that will connect beliefs and desires on the one hand and the actions we agree they cause on the other. Of course, in searching for such a law to fulfill its own theoretical needs, economic theory is also serving the ultimate needs of all those other social sciences that require nomological foundations for their singular causal claims but do not devote themselves self-consciously to the discovery of such laws. And no matter how much exponents of other social sciences may ridicule the findings of economics in its search for such laws, and proclaim either the disconfirmation of its laws on the basis of their findings, or the irrelevance of its claims for their projects, the singular claims of the political scientist, the sociologist, the historian, and even the anthropologist and ethnologist must ultimately remain inexplicable in the absence of laws of the sort that economics has searched for these last hundred years.

Thus, in an important respect, economics is the most fundamental of the social sciences, and the one which, superficially at any rate, seems most in accord with the concepts and structure of empiricism's picture of natural sciences. Economics seems clearly the most advanced of the social sciences in its degree of quantification, in the range of theoretical agreement on the part of its students, in the uniformity of its presentation, and in the problems which it treats as paramount.[3] Economics seems fundamental, not in the sense that other social theories might be reducible to economics in the way that empiricism claims other natural sciences are reducible to physics, but, from our perspective, in its self-conscious attempt to express and substantiate a general law of human action from which all economic behavior follows and from which, properly generalized, its proponents claim, all human behavior follows. In their assumption of a principle of rationality economists have traditionally kept one eye on the generalizability of this principle to the explanation of all human action, even while they themselves have employed it to explain only a very small range of human actions and their aggregations. It is in this respect that economic theory may be considered the most basic theory in social science, for it purports to begin with a nomological assumption about human behavior which is itself *noneconomic* and thereby not only provides ultimately noneconomic explanations of economic phenomena, but also the prospects for general explanation of all other social phenomena that reflect human action. Unfortunately, although economics may seem to deserve a specially fundamental place among the social sciences on the strength of its empiricist and reductionist proclivities, it is in another respect the least fundamental and also the least systematically advanced of the social sciences. For though it is replete with generalizations deduced from the law of human action it assumes, it is devoid of singular causal claims with anything like the empirical and observational warrant of other social sciences. That is, there are few if any singular claims about the actual causes and effects of economic events, great and small, on which economists are likely to agree. This singular state of affairs is reflected in the jest that if all the economists in the world were laid end to end they still would not come to a conclusion. It is an enduring embarrassment for empiricism that the one social science that seems most clearly and unanimously to guide itself by empiricist lights is no more practically useful, and considerably less in agreement about the singular facts to be explained, than, for instance, the least general of subjects, history. Empiricism had better have a plausible explanation for this state of affairs, as well as for the economist's failure to find a general law of human action, if it is to continue to make its prescriptive claims for social science. With this demand in mind let us turn to an examination of economic theory's attempt to provide the general law that all the social sciences require, and to the problems which prevent it from doing so. The examination will enable us not only to exclude a number of possible explanations but also to begin to suggest the outlines of an explanation of this failure that is both plausible and consistent with the continued enforcement of empiricist research strategies.

If individual actions are the causal consequences of the joint operation of desires and beliefs, then it seems obvious that the laws underlying this causal connection should appeal to desires and beliefs in their antecedents and describe actions in their consequents. Such universal conditions will be expressed at varying levels of generality, and it is to be expected that, like the general statements of natural science, at the lower levels of generality these expressions will be shot through with exceptions, while at the higher levels they will be connectable to actual cases only by a number of intervening auxiliary assumptions. Thus, the statement that 'wood floats while iron sinks' is of low generality and is false; similarly, the statement that 'everyone who wants to retrieve his hat and believes climbing a ladder will enable him to, does so' is also false and not the sort of generalization that is required to underwrite a wide variety of singular judgements, let alone all of them. The economist's notion of what the appropriate level of generality should be, of course, begins with the doctrines of Bentham and Mill on beliefs about the utilities of various alternatives, and the desire to maximize utility, although Mill's own economic treatise did not explicitly appeal to such forces in the form of a law to explain economic behavior. The explicit adumbration of a law of human behavior was left to the marginalist economists of the last third of the nineteenth century: Jevons, Walras, and Menger.

The marginalist economists built a theory of economic behavior on three distinct general claims: that all agents have complete information about all the alternatives facing them, i.e., they have true beliefs about all facts relevant to their circumstances; that they can rank these alternatives the circumstances provide in order of preferability by considering the cardinally measurable utility each will provide the agent; and that their desire to maximize their cardinally measurable utility combined with their belief that one among the alternative actions available to them will so maximize utility jointly determine the action they perform. Now the assumption of complete information is a boundary condition and is plainly unreasonable as a generalization; it was introduced in part only to simplify the construction of theory by enabling the generalization that agents choose those actions which they believe will satisfy their desire to maximize utility to be applied to idealized but nevertheless economically interesting situations. The marginalists did not treat their generalizations about the cardinal utilities of all objects, or events, for individual agents as a simplifying assumption; rather, they treated it as an empirical discovery of psychology. Similarly, they believed that as a matter of fact humans did act in accordance with the generalization that agents choose that alternative which they believe will satisfy their desire to maximize utility. Of course, they recognized that in the real world, by constrast with their special idealization, perfect information did not prevail, and that agents sometimes made mistakes. But these mistakes reflected mistaken beliefs about the availability or cardinally measurable utility of alternatives, and not the failure of the agent's beliefs and desires to determine his action.

The marginalist economists were perfectly clear in their view that in the notion of action as utility maximization they had hit upon a formal quantitative version of the general principle that underlies all distinctively human behavior. As Wicksteed put it in *The Common Sense of Political Economics*, "Our analysis has shown us that we administer our pecuniary resources on the same principles as those on which we conduct our lives generally. . . . In the course of our investigations we have discovered no special laws of the economic life, but have gained a clearer idea of what that life is." And further, "Every purchase being a virtual selection and involving a choice between alternatives is made in obedience to impulses and is guided by principles which are equally applicable to other acts of selection and choice. To understand them we must study the psychology of choice. . . . We are constantly weighing apparently heterogeneous objects and selecting between them according to the terms on which we can secure them."[4] The principle that action maximizes utility believed available is suitably general for the explanation of all human action, for it establishes one goal or desire as paramount and enables us to explain plausibly every other desire cited in a singular causal statement as a desire for some means to this end, or as a nongeneral description of this end, and to assimilate in a general way the factual beliefs about what alternatives are available for attaining this desire. The economists, however, restricted their employment of the principle to an explanation of the demand that agents have for various goods and services. To do this they added a further general psychological claim, that the utility of any item to any agent was a marginally decreasing function of its quantity, and were thus able to deduce that the amount of any good an agent chooses is a decreasing function of the price of that good.[5] But, of course, because of its generality the principle should be applicable to any human action the strength of whose antecedent desire and content of whose antecedent belief we can determine.

This unfortunately proves the Achilles heel of the principle, for we are unable to provide further generalizations which will enable us to measure the cardinal utility of any alternative facing an agent. And without such a further general claim we are unable to apply the principle to actual cases. It is easy to say that whatever the agent chooses is the alternative with the highest cardinal utility, but to so claim is to invoke the very explanatory principle whose applicability is in question in the establishment of its initial conditions of application. While it may seem initially plausible to measure the strength of different desires of a given individual on a scale which, like the cardinal weight scale, has a natural zero somehow corresponding to indifference to an outcome, it turns out that there are no natural units available to measure cardinal utility. This is just another way of saying there are no generalizations relating distinct amounts of cardinal utility to distinct amounts of something else which is itself directly or indirectly cardinally measurable. The required generalizations, some marginalists supposed, should be left to psychology, and in particular psychophysics, to discover, since utility is supposed to bear a relation to felt pleasures and satisfactions. Needless to say, this relation is too vague and speculative to be taken very seriously, but the notion that desires qua mental states must differ from each other quantitatively if they have differing effects on actions chosen is certainly a reasonable one and offers some justification for the cardinalist view of utilities. The fact that neither introspective, behavioral, cognitive, nor neurophysiological psychology has yet provided a theory which implies such quantitative differences between differing states of desire is the real reason why the cardinal notion of utility was surrendered and why the general lawlike claim that agents maximize utility so conceived was surrendered in the course of economics' development.

It has been fashionable for sixty years or more to describe the eclipse of cardinal utility as reflecting its operational meaninglessness, and to describe the advances in economic theory as reflecting a closer attention to empiricist strictures on the introduction of theoretical concepts. Once it was discovered that demand curves could be derived from assumptions about utility weaker than the marginalists' commitment to cardinality, the justification for surrendering the notion came to be largely methodological and to involve the suggestion that notions of cardinal utility were meaningless because they involved illegitimate reliance on unverifiable introspection. In fact, had economists and psychologists of the period discovered even the most *indirect* nomological connection between differences in the strengths of desires (which we know surely exist), and some naturally measurable differences in behavior or physiology, they would have retained the notion of cardinal utility despite its introspectionist flavor and operationally independent status. The failure of cardinal utility was simply an empirical one, and not a conceptual or methodological one. It just happens that there are no laws we can discover about cardinal utilities, and so no explanation of actual phenomena can appeal to them. The notion is legitimate and intelligible, but simply reflects a blind alley in the search for the *units* with which to measure one of the causal determinants of behavior. In this respect it was no worse than the pre-Mendelian individuation of hereditary properties; the pre-Mendelian classification schemes resulted in the misclassification of hereditary and nonhereditary traits, just because they hinged on the application of principles of heredity that were empirically false. Just as it took the discovery of the correct theory of heredity to provide the right categorical scheme for hereditary properties, the classificatory employment of a certain conception of utility requires the truth of a particular theory about utilities. If no such theory is correct, the associated classificatory scheme is bound to be empirically unsatisfactory.

Indications that cardinal utility does not characterize a nomologically systematizable notion came in the recognition that on its ordinary interpretation, which made its amounts for any individual a function of the commodities he chooses, this sort of utility is not arithmetically additive, but a function of the availability of other commodities, complements, substitutes, "superior" and "inferior" goods, and so on. Thus, the quantitative amount of utility provided by an additional unit of, say, bread will depend not on its physical properties alone, but also on the availability of complements like jam, or substitutes like potatoes, on the income of the agent, and on its status as a superior or inferior good in his schedule of purchases. In short, no arithmetically manipulable numbers can empirically be associated with a commodity that will express the degree of satisfaction its consumption invariably or even regularly provides.

The systematic sterility of this way of measuring the strength of desires, coupled with its theoretical superfluity for purposes of economics, led to the abandonment of cardinal utility for a weaker notion that required only a measurement of the ordinality of strength of desires. For it was shown that the downward-sloping demand curves that microeconomic theory requires can be derived from hypotheses to the effect that agents can order their preferences from greatest to least, although these desires may bear no intrinsically metrical relation to one another. The appeal to this weaker notion of ordinal utilities was first made by F. Y. Edgeworth and Wilfredo Pareto. One assumes, not that individuals attach a numerical quantity of desirability to alternatives, but simply that they rank for preferability each of the alternatives open to them in such a way that they can tell when they are indifferent between varying amounts of any two commodities available to them. The result of this ascription to individuals is the construction of curves of indifference for pairwise comparisons of all available alternatives, and multidimensional surfaces of indifference between amounts of all the commodities available to him. The agent is then assumed to choose that combination of alternative goods, from among those which he believes are available to him, which he most prefers—in the language of indifference curves and budget constraints, the combination of commodities uniquely specified by the interaction of the equation determining his budgetary limitations and the indifference surface furthest from the origin among those intersected.

The attribution of preference orders and their expression in indifference surfaces represent assumptions considerably weaker than the attribution of cardinal utility but still general enough to underwrite lawlike foundations for the explanation of economic behavior and seemingly extendable to the explanation of actions that are not obviously choices over spatiotemporally discrete goods and services. In particular, ordinal utility does not seem to suffer from the nomological and evidential isolation of cardinal utility, for we can measure it independently of the maximization hypothesis in which it functions, or so it seems. To determine an ordinal utility ranking, it is not necessary to find a quantitatively varying correlate in physiology or behavior; it seems enough simply to ask the agent to verbally disclose his preferences among alternatives stipulated by the questioner. And even when this is not practicable, it seems reasonable to infer from the counterfactual that the agent could express such a scheme of preferences, that he does have one. Naturally, our "measurement" of preference structures may not be very systematic nor even produce regularly replicable results, but the attribution of them to individuals on the basis of their actual behavior is as firmly established a feature of our ordinary explanatory apparatus as the assumption that reasons cause human behavior. Unfortunately, on the assumptions that individuals have such preference orders and that we can elicit them by constructing a questionnaire, the hypothesis that they always act so as to attain their most preferred objective, subject to their beliefs about availability of alternatives, turns out to be false. It may even turn out to be false in cases where it is reasonable to suppose that the beliefs about alternatives have not changed. In such cases it is often plausible to preserve the maximization hypothesis by hypothesizing so-called "exogenous" changes in tastes between the time the preference questionnaire was answered and the time the explained or predicted behavior was manifested.

But to secure the maximization of the ordinal preference hypothesis as something more than an unfalsifiable tautology in the light of these circumstances requires more than just our rough-and-ready counterfactual belief that had the questionnaire been applied at the moment of actual choice, it would have produced a map of the preferences which the actual choice reflects. In short, what is required is again a law, a nomological regularity, which will connect preference orders and some other variable that can be measured independently of actual choices made as a causal consequence of the preference orders. In the absence of such a regularity to legitimate preference order as the unit of the causal variable of desire, ordinal utility floats in as much of a systematic void as cardinal utility. It will not do to reply that ordinal utility is known to be a fruitful explanatory notion even in the absence of laws connecting it with other variables measurable independently of desire's causal consequences, because whatever arguments can be constructed for such a conclusion have their companion in arguments for cardinal utility as well. For all such arguments trade on the assumption that anecdotal behavioral evidence for preference rankings can be taken seriously because they reflect the operation of some as yet undiscovered law; but a similar argument will enable us to infer from the existence of another as yet equally unknown law that such evidence reflects cardinal differences as well as ordinal ones. If the operationalist rejection of cardinal utility is an expression of the failure to discover such laws, then it militates équally against ordinal utility measures.

Cardinal utility is often described as having been rejected because it was a notion that relied too heavily on introspection, since, because of its unaccessibility, introspection is illegitimate as a basis for the introduction of scientifically admissible notions. But in fact, the attribution of independent cardinal utilities was rejected because it appeared false on introspective grounds, and no alternative foundation could be provided for it. It was, after all, introspection that convinced economists that commodities' utilities were not independent of one another, and therefore not additive in the way required by the simple cardinal notion. Similarly, ordinal conceptions of utility were surrendered with the complaint that they still involved unnecessary or improper appeal to cognitive and conscious states of agents. After all, it was noted, the applications of maximization hypotheses that economics makes surely do not rest on agents' being conscious of and being able to accurately report their preferences; for these assumptions are immaterial to economic problems, and preference-order ascriptions should be interpreted in a way that reflects the irrelevance of these superfluous implications.

Thus the surrender of ordinal utility assumptions was described as a species of behaviorist revolution in economics. But the surrender actually occurred, not because a new, fashionable methodological imperative appeared, but because the reasonable behavioral assumption that ordinal preferences should systematically manifest themselves in behavior like verbal reports turns out to falsify or trivialize the lawlike claim that agents choose their most preferred alternative. And the later exclusion of desire as a causal variable in the explanation of human action, to which the surrender of ordinal preference eventually led, cannot be described as reflecting the influence of behaviorism in the explanation of human action, but rather as the surrender of any attempt to explain that sort of phenomenon at all. For the account of preference which superseded ordinalist accounts of it, the so-called revealed preference theory, preempts all questions about what desires causally determine individual actions. It does so because, to avoid crediting agents with introspectively available preferences that are publicly inaccessible and yet also undercut our confidence in maximization hypotheses, this theory derives downward-sloping demand curves from nothing but the minimal assumption that all agents' actual choices among pairs of alternatives (no matter how causally determined) are always transitive. If from this assumption about the agent's actual behavior, all the economically interesting consequences of traditional marginalist theory follow, then the contemporary economist will eagerly embrace the revealed preference approach; for it absolves him of the duty of finding ways of individuating and measuring the variables which he originally supposed to determine economic (and all other) action, and which he has been unable to isolate over the course of a century's efforts.

The price for shirking this task is that the economist is now no longer able to explain his original explanandum phenomenon: individual choice. Neoclassical economics began with the view, quoted from Wicksteed above, that its subject matter was but an aspect of the general activity of choice and was to be explained by appeal to laws of a psychological and not a narrowly economic nature. But if the most fundamental assumption economics makes about agents is that their actual behavior manifests a certain transitivity relation, then economics clearly forgoes the resources and the opportunity to explain this behavior, to explain actual choices. If the theory begins with actual behavior, it is neutral with respect to any theory or competing theories about the causal determinants of that behavior, and provides no explanation of it. It is perhaps paradoxical that the course of attempts to formulate successively more scientifically respectable and empirically warrantable versions of a principle to explain human action in terms of its causal determinants should end in the abject surrender of any pretensions to explaining these sorts of events. To parade this turn of events as the harkening of economic theory to the operationalist dicta of modern scientific method and its associated philosophy of science, as many economists have done, is fundamentally to misrepresent the history of a failed research program by changing the aims of the program after the failure has become apparent.

For thirty years after the work of Jevons, Walras, and Menger, cardinal utility was applied to a formalization of what was proposed as a literally true theory of individual choice and its consequence for economic aggregates. After the failure to provide independent specification and measurement for the explanatory variable, it was surrendered in favor of a weaker notion, which held sway for nearly another forty years, though it too suffered from failure, this time of a more complex sort: for the natural interpretation of ordinal preferences turns out to falsify the maximization hypothesis with which it was associated, except when that hypothesis is protected by appeal to indeterminable changes in the exogenous variable of taste. And so, instead of either surrendering the maximization hypothesis or continuing to search for an account of preferences that would enable us to apply and confirm the hypothesis, economics surrendered the object of explaining individual economic behavior. It did so in two ways. First, it embraced revealed preference theory as the cornerstone of its theory; and second, the more sophisticated of its theorists began to deny the appropriateness and applicability of the traditional interpretation of its formalism in terms of beliefs and desires of individual agents. They did so by pretending that all along economic theory had no interest in the behavior of the individual agent and that its apparent reference to his beliefs and desires was but a *façon de parler*, a fiction for computational convenience, in the attempt to systematize the real subject matter of economics: markets and economies as a whole. This tradition, which goes back to Marshall, has its most impressive exposition in one of the classic accounts of ordinal utility theory. In *Value and Capital*, Sir John Hicks writes: "In our discussions of [the law of consumer demand] we have been concerned with the behavior of a single individual. But economics is not in the end much interested in the behavior of single individuals. Its concern is with the behavior of groups. A study of individual demand is only a means to the study of market demand." Elsewhere he notes that since "our study of the individual consumer is only a step towards the study of a group of consumers . . . falsifications may be trusted to disappear when the individual account is aggregated."[6] This attitude, tenable or not, marks a clear break with earlier economists in Hicks' tradition; more important, it in no way absolves the economist of the employment of notions like preference, for which he can give no account compatible with the truth of his leading explanatory principle.

Of course, experimental psychologists were originally attracted by their own interest in the explanation of choice behavior to the examination of the economist's explanatory variables, and to his own particular treatment of them. Earliest among them was the distinguished experimentalist L. L. Thurstone, who in the twenties attempted to infer experimentally indifference maps for subjects from behavior and to predict subsequent indifference maps on their bases.[7] Thurstone's attribution to his (single) experimental subject of an indifference curve between hats and overcoats, hats and shoes, and shoes and overcoats, enabled him to predict an indifference curve for his subject between shoes and overcoats. The method involved assuming that the subject maximizes utility and that utilities of different commodities are independent (a patent falsehood in the present case), and constructing from preference behavior data utility curves for shoes and overcoats separately, and then summing them for different combinations of each. Several things are worth noting about this apparently unique experimental attempt to individuate and quantify desires of the sort that the economist and other social scientists study. First, as Thurstone himself noted, the subject's consistency of choice was remarkable, and he attributed this to the experimenter's careful instruction "to assume a uniform motivational attitude." Secondly, in an attempt to acquire behaviorally based access to paired preferences, the experimenter required appeal to utilities, which are cardinal to the extent that their amounts are assumed to be additive and independent of one another. It is odd that a notion whose behaviorist virtues economists extolled required a nonbehaviorist notion for its isolation and description. This being the case, the special methodological merits of ordinal utility turn out to be illusory. Finally, the assumptions of utility maximization and of utility independence may be claimed to rule out Thurstone's method of measuring preferences, as one which will enable us to assess the merits of the economist's hypothesis of utility maximization; for it clearly begs this very question. I say that this charge may be laid, but not that it must be, or will be, decisive, because all scientific measurement involves a circle of notions and assumptions, and the real questions about their adequacy are whether the circles are large enough to provide their elements with explanatory power. This is an issue to which, in the present connection, we will turn in the next chapter.

But the main limitation on Thurstone's experimental technique for constructing indifference curves—aside from the difficulty of generalizing them from paired comparisons to n-tuples of the dimensions agents actually face, and aside from the fact that in the absence of careful instructions, preference behavior turns out to be inconsistent and insensitive—is that it enables us at best only to predict the subject's indifference curve for choices between combinations of commodities already offered in conditions of certainty. That is, it enables us to predict no actual choices among these commodities in conditions of uncertainty. It is surely no objection to Thurstone's method that his experiment was closed to the effects of uncertainty in the agent's beliefs, for it is only reasonable in attempting to find a behavioral measure for one psychological variable that it should be allowed to vary in circumstances from which another codeterminant of the same behavior is excluded. The problem is, when we come to measure the second variable, belief, whether we can hold desire constant in the same way. If we cannot, our confidence in the ability of these two variables to explain systematically and precisely the consequences we accord them will be seriously undermined.

Closing the explanandum phenomena to the effects of variations in belief has been a central characteristic of economics since the time of the marginalists. The ubiquitous assumption that all agents have perfect information about available alternatives is nothing less than the attempt to insulate the explanation of consumer behavior against the influences of doxastic states of varying degrees of strength, as well as different contents, and with varying degrees of influence on behavior. And just as such insulation was reasonable for psychologists like Thurstone, it was equally legitimate during the period in which economists were attempting to clarify the identifying marks of variations in the strength of preferences in behavior. Naturally, it is reasonable to expect that once ways of measuring each of these distinct causal forces are discovered, we will be able to bring them together in tests of the claim that they are jointly sufficient for the nomological determination of human action. In effect, our expectation is identical to the expectation, eventually fulfilled, that measurements of pressure and temperature of a gas, independent of each other, and of volume of the gas, could both be employed to establish that volume is a nomological function of pressure and temperature.

Although economists never did solve the problem of measuring strength of desire independent of behavior, and psychologists likewise made little progress on the problem, both recognized the importance of providing measures of strength and content of beliefs, for both eventually began to focus on the problems of decision under conditions of imperfect information. Of course, neither economists nor psychologists expressed the problem of quantifying the causal variable of belief in this way; rather, they set themselves the problem of waiving the assumption of perfect information and explaining human action under conditions of uncertainty and risk. "Risk" is generally used to describe those cases in which our beliefs about alternatives available can be given a probabilistic measure that reflects our degree of confidence that these alternatives will actually be available. "Uncertainty" is often used to describe situations in which our knowledge of the availability of alternatives does not extend even to the attachment of probabilities; these are situations in which we have almost no knowledge about availability of alternatives. The most important theory of decision under conditions of risk is of course that of John Von Neumann and Oscar Morgenstern.[8] In effect, they showed that it is possible to construct a utility measure which, together with the maximization of utility hypothesis, determines agents' actions in the face of alternatives to which they can attach probabilities. Determinate choices under conditions of assessable risk follow from five axioms about agents' preferences among alternatives available with certainty and/or with varying degrees of probability as well as among combinations of alternatives available with certainty or probability. Now, while this model represents a formulation of the traditional economist's maximization hypothesis that enables us to apply it to the explanation of action in the absence of certainty of belief about outcomes and is therefore much more relevant to the explanation of actual behavior than the original hypothesis, it not only does not avoid the specification problem for preferences that bedeviled the marginalists, but it also generates a specification problem for the probabilistic beliefs it accords to agents. In the absence of such a specification, actual predictions of choice under conditions of risk cannot be made because we cannot measure the probabilities which agents assign to alternatives which, in their beliefs, may be available. Of course, we can infer their beliefs from their actual choices, on the assumption that their preferences satisfy the five axioms of the Von Neumann-Morgenstern theory, and that they maximize the so-called "expected" utility which that theory accords agents, but such an inference will not allow the independent specification of probability assignments that we require. What is worse, the axioms seem independently to be falsified by the apparent fact that some agents derive pleasure from the sheer opportunity to take risks, even at losing odds, and others avoid risk at any cost. Such agents violate the Von Neumann-Morgenstern axioms because they are never indifferent in their preferences between a given alternative with certainty and a combination of more and less preferred alternatives with given probabilities. The axioms seem to be falsified by apparent risk avoidance and risk attraction, but we really do not know whether they are or not because we have no independent measure of either preferences or subjective assessment of probabilities, which we require in order to apply and test the Von Neumann-Morgenstern account.

Just as the economist's problem of specifying preferences was reflected in experimental work of psychologists, similarly, the notion of subjective probability as a measure of strength of belief also attracted the attention of psychologists. It has, however, been an invariable feature of all the sustained accounts of subjective probability attributions that they all rest on a prior attribution of some preference order or utility structure to their subjects. That is, each involves the employment of actual behavior as a measure of strength of belief, but only on the assumption that the experimental subject has, for example, a specified utility scale for money, or for some other reward or reinforcement. A number of independent experiments performed in the late forties and fifties[9] showed considerable agreement on the question of how subjective estimates of probability vary with objective, actual probabilities; but each involved the assumption that money payoffs determine a preference structure, and so none provide a measure of strength of belief independent of assumptions about the existence and causal force of varying strengths of desire. The potentially inextricable connection between belief and desire is reflected in work by philosophers and mathematicians on the foundations and interpretation of probability and statistical theory. Thus, for instance, Frank Ramsey's influential paper "Truth and Probability" involves explicit appeal to utilities in the measurement of probabilities; and L. J. Savage's *Foundations of Statistics* propounds a measure of probability for all events on the assumption, among others, that all actions can be ordinally ranked for preference.[10] Of course, there seems to be an alternative in the determination of subjective probabilities to complex inferences from choice behavior that presupposes preferences: just ask the subjects to communicate their strength of belief by making verbal estimates of probabilities. This method is akin to the notion that we can measure preference strength independent of behavior simply by asking the subject to verbally rank available alternatives. Experiments have been performed to test hypotheses about the relationship between subject measures of probability and actual objective probabilities simply by asking subjects to estimate the frequencies which are known independently by the experimenter. The general conclusion has been that subjective measures of probability are roughly a linear function of objective probabilities, although not identical to them, and that they vary over individuals.[11] The trouble with such specifications of strength of belief is identical to the problems associated with verbal reporting as a measure of strength of preference. If accepted, they show maximization of utility or preference hypotheses to be false. On the other hand, economists and psychologists do not hold that the applicability and descriptive accuracy of a maximization hypothesis rests on agents' abilities to be continually conscious of or able to report accurately the estimates of probability governing their actions. Thus, direct questioning is rejected as a source of identification for one of the causal determinants of behavior on the grounds that it makes methodologically illegitimate appeal to phenomena that are only introspectively available and cannot be publicly confirmed.

Economists seem to be faced with the following difficulty. Their employment of any lawlike claim to the effect that actions reflect the maximization of preferences, subject to beliefs about alternatives, requires a specification of the beliefs and desires in question. And yet none of the specifications of desires that trades on our commonsense understanding of this notion seems to confirm the maximization hypothesis, and what is worse, no specification of strength of belief seems available in which that notion is isolated in the required way from the notion of strength of desire. Accordingly, the maximization principle can be preserved only on the condition that it is not actually applicable to the practical prediction and control of any particular human action, for we do not have the resources to establish the initial conditions for such predictions. In the absence of the required auxiliary hypotheses, economists found a number of ways to preserve the employment of maximization hypotheses. One of these ways involved using behaviorist and operationalist methodological dicta to exclude interpretations and specifications of belief and desire that disconfirm the hypothesis. Sometimes this tactic has been joined to the proclamation that economic theory is not to be understood as dealing with the actual behavior of individual agents. The trouble with such a proclamation is at least twofold. First, it is rejected by many economists and belied in their actual work, so much so that it seems more a rationalization for their failure to find the required specifications than an expression of the aims and claims of economics. Second, and more important, in searching for a general law relating action with reasons, economic science has been doing duty for all the other sciences that trade in these same causal variables and yet do not explicitly formulate the law which a commitment to empiricism requires to underlie them. Insofar as the other social sciences actually do purport to provide explanations of individual human action by appeal to beliefs and desires, they can hardly adopt the economist's disdain for individual human action as a fitting subject of study.

Indeed, the surprising feature of developments in the other social sciences during the period in which economists increasingly lost faith in their theory as a vehicle for the explanation of individual human action was the growth in the influence and emulation of economics and the explicit appeal to a maximization hypothesis by these other subjects. Thus, the attribution to agents of ordinal preferences, of beliefs about probabilities, and of the interaction of these in the determination of actions in accordance with varying formulations of the maximization hypothesis have become the stock-in-trade of a host of theories in social psychology, in sociology, and, most apparent of all, in political science. The explanation for this expansion of *Homo economicus*'s influence is to be found partly in the mathematically impressive formalism that economics has managed to erect on the maximization hypothesis, and partly in the increasing recognition that the singular causal claims of these subjects require the sort of systematization and foundation that only a lawlike generalization can provide. Social scientists are not about to give up the claim that action is the product of belief and desire, but they increasingly recognize that their accounts of action need to reflect certain features hitherto common in natural science: their findings need to be expressed quantitatively and interconnected deductively. In short, social scientists are coming to recognize what empiricists could have told them all along: that they require laws connecting beliefs and desires with actions. The empiricist commitment to the existence of laws and the common assumption that human behavior is the effect of individual reasons are simply too strong to have been surrendered in the face of failures over the course of a mere century to provide the means of expressing, testing, or applying laws of the required form to actual cases. Thus, despite the vagaries of fortune and fashion which have shaped both the economist's view of his own theory and other social scientists' assessment of its extendability to their own subjects, the hypothesis of maximization finds itself in the same fundamental position that Wicksteed accorded it in 1910; for, as he wrote, so it is still believed, that "we administer our pecuniary resources on the same principles as those on which we conduct our lives generally."

But empiricism about reasons, which leads to the search for laws of human action, cannot in good conscience simply ignore the problem of independent specification of belief and desire, nor can it indefinitely close its explananda to uncertainty or incompleteness of information. And indeed, the best of current social science, in its continued commitment to empiricism about reasons, attempts to circumvent the difficulties which have faced earlier endeavors. It is important to examine at least one such attempt in detail, both to determine whether these problems have been superseded and to reveal in a new way the force and the potential ubiquity of the economist's strategy in the explanation of all human behavior. The upshot of our examination will be the paradoxical conclusion that thoroughgoing empiricism about reasons forces us to choose between a continuing commitment to empiricism or to reasons as the *nomological* determinants of action.

The attempt to propound a unified theory of all human behavior that we will examine is that of Gary S. Becker. Becker's work, spread over a twenty-year period from the middle fifties onward, represents a sustained, detailed, and novel argument for the claim that "the economic approach provides a valuable unified framework for understanding *all* human behavior." In the course of a career devoted to the establishment of this claim, Becker has provided an extended and systematic application of the model of rational economic man, *Homo economicus*, to a whole host of problems previously limited to unsystematic treatment by sociologists, criminologists, anthropologists, behavioral and developmental psychologists, sociobiologists, and others. But his application is specifically made with a view to avoiding many of the objections traditionally made against maximizing models, and some of the defects canvassed in this chapter. In a sense Becker's work represents something of a *best case* for the explanatory powers of *Homo economicus*, and thereby, of the strength of the commitment to empiricism about reasons, that we are likely to see in the near future. In 1976 Becker brought much of his work towards the completion of his program together in a single volume, *The Economic Approach to Human Behavior*.[12] In his introduction Becker writes that he has "come to the conclusion that the economic approach is a comprehensive one that is applicable to all human behavior, be it behavior involving money prices, or imputed shadow prices, repeated or infrequent decisions, large or minor decisions, emotional or mechanical ends, rich or poor persons, men or women, adults or children, brilliant or stupid persons, patients or therapists, businessmen or politicians, teachers or students" (p. 8). Naturally, if empiricism holds sway, then anyone who seeks to explain human behavior as action is committed to some version of the "economic approach." For Becker the economic approach means something quite specific: it consists in three basic features: "the economic approach assumes (1) maximizing behavior more explicitly and extensively than other approaches do, be it utility or wealth functions of the household, firm, union, or government bureau that is maximized. Moreover the economic approach assumes (2) the existence of markets that with varying degrees of efficiency coordinate the actions of different participants . . . so that their behavior becomes mutually consistent. . . . Prices and other market instruments allocate the scarce resources within a society and thereby constrain the desires of participants and coordinate their actions. In the economic approach, these market instruments perform most, if not all, of the functions assigned to 'structure' in sociological theories." Most important, the economic approach assumes (3) preferences "not to change substantially over time, nor to be very different between wealthy and poor persons, or even between persons in different societies and cultures" (p. 8). It attributes one and only one preference order to every human being that has ever lived or will live.

This last assumption is crucial to Becker's theory in at least two respects. By making it he precludes any defense of the maximization model against apparent falsifying experiments by appeal to changes in taste, and at the same time he obligates himself to provide a theory of maximizing behavior that explains why different agents with apparently the same available alternatives often choose differing options. A conventional theory of rational choice can always be preserved in the face of a false prediction about individual choice behavior by the claim that its auxiliary hypotheses about the preferences of the agent observed have changed during the test period. The fact that defenders of the conventional theory so frequently appeal to this defense, coupled with the fact that there is no means of determining preferences independent of actual choices, frequently leads to the charge that the theory of rational choice is unfalsifiable. By assuming stable preferences Becker simultaneously forecloses this method of defending his version of rational choice theory and avoids the charge that it is unfalsifiable on this ground. More significant, he obviates the need for principles which will individuate and quantify preferences in each and every case of human action. On the other hand, since agents faced with the same alternatives almost invariably make different choices (for example, from a menu or a dress rack, among investment and insurance alternatives, colleges, careers, etc.), the assumption that preferences are the same both across time and between agents, no matter how culturally and socially different, cannot be coupled with traditional theory to explain any part of this tremendous variation in choice behavior. The assumption, together with conventional theory, predicts that we all make the same choices, and this is patently absurd. Accordingly, Becker is committed to producing a new theory or a new version of the old theory of rational choice if he is to render consistent the stability of tastes and the variation of choices. The theory which he produces offers one of the strongest arguments for his general claim.

Of course, if preferences are stable in the way that Becker requires, then they must be very different sorts of things from the conventional picture of them. They cannot be preferences for one brand of cigarettes or one flavor of ice cream, since not only is there no interagent stability in these preferences in our culture, but there is even less comparability among preferences of this sort between agents in our culture and an agent in the cultural setting of ancient China, for example. As Becker recognizes, "Preferences will have to be defined over fundamental aspects of life, such as health, prestige, sensual pleasure, benevolence, envy, that do not always bear a stable relation to market goods and services" (p. 5). It is highly plausible to assume that preferences thus characterized are stable across lives and individuals and can be specified in, for instance, biological generalizations. But how can such general and indubitable preferences be connected clearly and precisely to those specific, mundane, and incredibly various choices that make up so much of the texture of every individual's daily life? Here again a new theory is called for, one which not only can make no appeal to changes in taste, but which also cannot write off very much human behavior as *irrational*, in the face of these stable preferences. For to do so is to surrender Becker's pretension to explain all human behavior, as well as to open the theory to the charge of unfalsifiability from another quarter. To provide the connection and to avoid the charge, Becker's approach

> does not assume that all participants in any market necessarily have complete information or engage in costless transactions. Incomplete information or costless transactions should not . . . be confused with irrational or volatile behavior. . . . The assumption that information is often

> seriously incomplete because it is costly to acquire is used in the economic
> approach to explain the same kind of behavior that is explained by irra-
> tional and volatile behavior, or traditional behavior. . . . [Thus] when an
> apparently profitable opportunity . . . is not exploited, the economic
> approach does not take refuge in assertions about irrationality, content-
> ment with wealth already acquired, or convenient ad hoc shifts in . . .
> preferences. Rather it postulates the existence of costs, monetary or
> psychic, of taking advantage of these opportunities that eliminate their
> profitability (pp. 6-7).

It is to the variation in available information and the costs of its acquisition
that Becker's theory ultimately appeals in order to explain how stable preferences give rise to diverse choices.

But what of the simple and appealing objection touched on above in connection with traditional economic theory, that human agents simply do not
behave on the basis of decisions taken in terms expressible by the differential
calculus of microeconomics, that they do not in fact actually literally *calcu-
late* which among an exhaustive list of consciously recognized alternatives
will quantitatively maximize some psychological variable like preference or
utility. It was this sort of objection, I suggested, above, that initially led the
economists following Marshall to disclaim any real interest in explaining
actual individual choice, and provides an important motivation for the operationalist interpretation of choice theory in terms of revealed preference.
For on its behavioral interpretation, rational choice theory is not open to this
sort of objection, though at the cost (which its exponents seemed willing to
bear) of being inapplicable to the explanation of actual individual choices.
Becker, however, cannot avail himself of the advantages of the revealed preference approach to rational choice theory, for his aim is to explain all behavior as the making of choices determined by preferences, and not to treat
preferences as *façon de parler* for actual choices. How then can he avoid the
charge that he employs a characterization of decision processes that is incompatible with the findings of psychology about cognition and patently false
from introspective considerations alone? To this complaint Becker's answer
is that "the economic approach does not assume that decisions [*sic*] units
are necessarily conscious of their efforts to maximize or can verbalize or
otherwise describe in an informative way reasons for the systematic patterns
in their behavior." Becker goes on to describe this assumption as "consistent
with the emphasis on the subconscious in modern psychology and with the
distinction between manifest and latent functions in sociology" (p. 7). This
"black box" approach to the nature of the cognitive psychological theory
that might underlie a general account of human behavior is identical to the
conventional theorist's refusal to provide an interpretation or specification of
beliefs and desires and as such leaves open serious questions about the nature
of human agents, among which is the question whether, in the light of

Becker's denial of any uniform conscious mechanism associated with rational action, they really have any claim to be called *agents* at all. We will return to this matter later. Let us now turn to the details of how Becker translates these claims about "the economic approach" into a theory of human behavior.

According to the traditional theory of rational choice, in its original economic setting, the consumer attempts to maximize utility, U, which he obtains from goods and services, x_i, purchased in the marketplace, subject to the constraint of his income. How much of each good he chooses—his demand function—is determined by his income, by the price of goods in the market, and of course by his tastes, reflected in his utility function. These relationships are typically expressed in the following simple equations, for utility function:

$$U = U(x_1, x_2, \ldots, x_n)$$

for income constraint:

$$I = \sum_{i=1}^{n} x_i P_i$$

and for demand function:

$$x_i = d_i\left(\frac{I}{p}, \frac{P_n}{p}, \frac{P_i}{p}, T\right),$$

where U = utility, x_i = quantity of market good i chosen, I = income, P_i = price of good i chosen, p is a price index, and T = tastes. As the demand function makes obvious, on this view, only three things determine actual choices: income, prices, and tastes. As noted above, this formulation suffers from a number of defects, not the least of which is the theory's reliance on differences in tastes to explain differences in choice, when "economists have no useful theory of the formation of tastes," and can rely on no "well developed theory of tastes from any other discipline" (p. 133). In other words, it has no account of preferences independent of the behavior preferences are cited to explain. Moreover, "by implying that utility is derived from goods and services purchased in the market place, the received theory has generally been formulated in terms of monetary prices and monetary income. . . . This concentration on analyzing responses to monetary phenomena has considerably limited the theory's appeal to other social scientists. The political scientist, sociologist, or anthropologist is typically concerned with behavior where monetary phenomena are not pervasive. . . . Small wonder when that theory relies so heavily on money prices and attributes so much of observed behavior to unexplained variations in taste" (pp. 133-34).

Suppose that instead of making the agent's utility a direct function of goods and services he purchases on the market, we assume that his utility is derived more indirectly. Suppose it is derived from some productive activity on the part of the agent (or the household of which he is a member) in which goods and services available on the market figure as factors of production, along with the agent's own time. Thus, the consumer's demand for market goods is "a derived demand analogous to the derived demand by a firm for any factor of production" (p. 134). This apparently innocuous variation in the characterization of the relation between available market goods and the agent's utility function immediately and substantially enhances the power of the theory, while avoiding the specification problems with respect to tastes and beliefs that bedevil traditional theory. To see this, we set out a small portion of the formalism of the theory.

The individual agent's utility function is

$$U = u\,(Z_1, Z_2, \ldots, Z_n),$$

where Z_i is not the quantity of a market good but the quantity of one produced by the agent for his own consumption . This commodity, Z_i, is produced by the agent by the use of quantities of goods, x_i, available in the market together with quantities of his own available time, t_i. Formally,

$$Z_i = z_i\,(x_i, t_i, E),$$

where E is a set of "variables which represent the environment in which the production takes place. These environmental variables reflect the state of the art of production, or the level of technology of the production process [that the particular agent employs to produce Z_i]" (p. 135).

This utility function is maximized by the agent subject to the usual income constraint expressed above, and also to a constraint on his available time:

$$T = t_w + \sum_{i=1}^{n} t_i,$$

where t_w and t_i are the agent's time spent in the labor market and in producing the nonmarket commodities, Z_i, which make for his utility.

Unlike the conventional model—in which the only determinants of actual choices are income, prices, and tastes—this model allows tastes to drop out altogether, and in addition to income and prices for market goods, its determinants include total available time for labor and for production of household goods (the Z_i's), and the nature of the productive processes the agent employs to convert market goods and time into household goods. And this last variable reflects all of those components of each individual's ability to get the most out of what is available to him; for the productive processes each agent employs in the generation of household goods will be a function of market good prices, available time, his own information about alternative production techniques, and so on. The mere multiplication of variables increases the prima facie explanatory power of a theory of rational choice. For now we can, for example, explain why two individuals facing the same market prices, and with the same income, choose different things, *without having to appeal to exogenous tastes*, about which we can say nothing. They may do so because of differences in available time, or information about productive processes, even though ultimately they have exactly the same tastes. Similarly, we may explain why education, or change in marital status, or any information-generating experience, may change the choices of an individual agent, even though his income and the prices of market commodities and even his tastes have remained unchanged. In fact, on this formulation the theory of rational choice will enable us to explain changes in what, on the received view, are unexplainable differences of tastes and preferences for market commodities by appeal to changes in productive processes for household goods. Changes in information about alternative ways of producing the same set of household goods will affect the sorts of and quantities of market goods desired, and in effect, impose changes on their so-called "shadow prices," which reflect their marginal productivity with respect to household goods. Thus, even when market prices remain unchanged, a change in production processes will change shadow prices of market goods for the agent and thus change his rate of consumption of them, even though the other variables of the traditional theory—income and tastes—remain unchanged. In short, the new theory has more prima facie explanatory power than the old one if only because "behavior differences previously attributed to differences in tastes are in fact due to differences in productive efficiency" (p. 145) and in the value of time available to combine with market goods in the production of household commodities. Since remuneration in the labor market for units of time varies over individuals and over the life cycle of a given agent, the amount of time distributed to the labor market (for the acquisition of market goods) and to household production will consequently vary, affecting the relative efficiencies of various production methods for producing the stable bundle of household commodities. These facts, too, may be formally expressed in terms of the shadow price of units of time, and both shadow prices for time and market commodities may be absorbed into a conventional theory of the marginal productivity of the factors of production which mathematically unifies the interrelations of these variables and their effect on actual choices in the market place and the labor market.

There are several respects in which this new theory does not really represent so vast a departure from the received one. The new theory is formally compatible with the conventional model in respect to individual responses to changes in relative price and income. Moreover, as Becker argues, "In the standard theory all consumers behave similarly in the sense that they all maximize the same thing—utility or satisfaction. It is only a further extension then to argue that they all derive that utility from the same 'basic pleasures' or preference function, and differ only in their ability to produce these 'pleasures.' From this point of view, the Latin expression *de gustibus non est disputandum* suggests not so much that it is impossible to resolve disputes arising from difference in tastes but rather that in fact no such disputes arise!" (p. 145). Although Becker's assumption of fixed and uniform tastes may represent a novel and useful extension of traditional theory, some may object that his new approach really represents no substantial departure from the theory as originally presented, or at any rate that it manifests methodological and factual limitations no less debilitating than the received view's reliance on tastes, whose nature and change it cannot account for. Becker himself moots the objection: "This shift in emphasis towards changes in prices and income and away from changes in tastes may appear to be simply one of semantics—of hiding an inability to explain tastes behind the camouflage of a production function" (p. 144). A philosopher will quarrel with the use of the expression "semantic" here, but the concern which this remark expresses warrants response. The charge is that all of the problems that the traditional theory sweeps into the variable for tastes, T, in its demand function for market goods, are swept into the "environmental variable," E, in the production function for household variables of the new theory. Just as the old theory tells us formally nothing more about tastes besides the fact that they affect the quantities of market goods demanded by an agent, the new theory's formalism apparently tells us nothing more specifically about these crucial environmental variables than that they determine the shadow prices of the same goods, and thus the quantities of them demanded by the individual agent. In both theories, the crucial explanatory variables are characterized and individuated by the very events whose occurrence they are cited to explain. Unless, in the extension of Becker's theory, independent characterizations of the variable E can be produced, both defenders of the old theory and detractors as well might complain that his alternative is as closed to falsification and as theoretically sterile as the received view turns out to be. But even in the absence of such detailed development of Becker's theory, the force of this objection is greatly reduced by important disanalogies between T and E, between tastes and environmental variables, even if at the present stage of development these two sorts of variables are methodologically on the same footing. As we have seen, the course of the development of traditional economic theory, from the time of Jevons at least, reflects the ever-increasing strength of the economist's belief that he has nothing useful to say about tastes and preferences, but can only discuss their effects. The history of economics reflects transition from a conception of utility that was originally cardinally ordered and interpersonally comparable, through several stages, until today standard presentations offer a conception of utility that is neither. So firmly and widely is this belief held by economists that more often than not they appeal to philosophical and not empirical arguments to sustain it. In short, economic theory holds out no hope of providing an account of the variable T, and makes a "value-free" virtue out of what it treats as a methodological necessity. By contrast, "economists profess to know something about factors associated with productive efficiency and have successfully studied such factors" (p. 144). Although we have no detailed general theory about the factors that determine E, the environmental variable, we can confidently make a large number of highly plausible and widely agreed-upon singular claims about the causes and effects of individual agents coming to use varying methods of producing those household goods that maximize their utilities. A theory of the determinants of productive efficiency and a basis for comparative judgements of such efficiency is clearly within the realm of empirical possibility with respect to the behavior of the firm. Indeed, the areas to which linear and dynamic programming have been applied show the existence of such a theory of productive efficiency to be more than a bare empirical possibility. What Becker's new theory of the consumer requires is a theory no different in kind from this one. Just because no such theory is as yet available is no reason to claim that the factors to which his theory appeals have no advantage over the admittedly exogenous explanatory variables of traditional theories of rational choice.

Chief among the virtues of Becker's new theory of consumer behavior is prima facie explanatory power: we may deduce from this theory a wider variety of different consumer choices than from the traditional theory of rational choice. In addition to implying statements about the direction of changes in amounts chosen by consumers among market goods as a function of changes in income, tastes, or prices, this new theory implies statements about changes (and their direction) even when these latter variables remain unchanged. Obviously it does so by introducing additional factors that do not figure in the traditional theory. Moreover, the new theory has formally greater explanatory power than the old, insofar as we may deduce from it and thus explain the preference-ordering among market goods which the traditional theory assumes as given and cannot explain. In short, with a little deductive manipulation we can show that the conventional theory of consumer behavior is reducible, in the logical empiricist's sense, to Becker's new theory.

But, of course, Becker's aim is not simply to produce a more general version of the economic theory of consumer behavior. His aims are much broader. In *The Economic Approach to Human Behavior* he applies this broadened theory of rational choice to a wide variety of human actions and far beyond traditional areas of economic interest. But even the breadth of his applications belies the scope of his claims. For the impressive range of phenomena which he purports to explain by his extension of traditional theory represents at best a set of examples or case studies which substantiate his claim to universal applicability. And although compatibility with a long but finite list of findings is all that Becker can show for his claim, we cannot help but be impressed with their range and detail.

Becker begins by applying the traditional theory of market behavior to the phenomenon of economic racial discrimination, showing that its existence in a free-market economy can be generated simply by strong tastes for discrimination among a majority, without the existence of political discrimination or concentrations of economic and political power. Next he constructs a theory of the conditions under which criminal behavior will be engaged in and of the types of crimes an agent is committed to the commission of by his beliefs and preferences. He shows that the theory constructed is consistent with reliable evidence on the incidence of crimes, rates of apprehension, and severity of punishment, and concludes that "a useful theory of criminal behavior can dispense with special theories of anomie, psychological inadequacies, or inheritance of special traits and simply extend the economist's usual analysis of choice" (p. 40). But it is only after introducing his new theory that Becker begins to expand the domain of the economic approach extensively. First a theory of the allocation of time by individuals between various activities is offered. The special importance of this treatment is its account not only of the determinants of the amount of time offered by agents on a labor market, but also of the allocation of time among nonmarket activities, where the traditional theory cannot function for lack of a price structure among these alternatives. Thus, we may explain, for instance, cross-cultural differences with regard to distribution of nonlabor time between, say, time-consuming rituals of food preparation and rapid eating, between the promptness and clock-watching of North Americans and the relaxed attitude towards tardiness in other societies, by appeal to the shadow price of units of the agent's time as determined by labor-market earnings, market prices, available processes for producing household goods, and the same stable household commodity preferences attributed to each individual agent, regardless of cultural setting.

More strikingly, Becker then proposes to explain the ubiquity of marriage in all societies; the sorting of mates for similarity in beauty, intelligence, education, wage rates, height, race, and religion; the incidence of polyandry, polygamy, and monogomy; the effects of love and caring on marriage; when and for how long to marry and remarry; and so on. All this Becker claims to do on the basis of two "simple principles": "The first is that since marriage is practically always voluntary, either by the persons marrying (or their parents) the theory of preference can be readily applied, and persons marrying (or their parents) can be assumed to expect to raise their utility level above what it would be were they to remain single. The second is that, since many men and women compete as they seek mates, a market in marriages can be presumed to exist. Each person tries to find the best mate, subject to the restrictions imposed by market conditions" (p. 206). Marriage, it turns out, enables the partners to acquire inputs and employ household production processes for the satisfaction of their utilities more efficiently than any other social arrangement. But since agents can join together in male-male or female-female pairs, or for that matter, in efficient households of larger sizes and heterogeneous proportions, the near universality of male-female pair unions cannot be explained simply by appeal to economies of scale in the production of household goods: "The obvious explanation for marriage between men and women lies in the desire to raise their own children and in the physical and emotional attraction between the sexes" (p. 210). The production of household goods that satisfy these two sorts of stable and univeral desires provides the fundamental motive for all marital choices, in Becker's view, and he traces out their consequences for all of the explananda indicated above.

Not only does Becker's account seem to be consistent with antecedently available data on marriage and mate sorting, but he also reports tentative *new* findings that confirm surprising and counterintuitive conclusions of his theory. Similarly, Becker applies his theory to the explanation of available census materials on national fertility rates and demographic patterns, showing that the aggregate trends these data report follow deductively from the attribution of the new model of rational choice to individual agents, and the treatment of children "as a durable good, primarily a consumer's durable, which yields income, primarily psychic income, to parents. Fertility is determined by income, child cost, knowledge, uncertainty, and the quality of children demanded" (p. 193). Becker recognizes that "it may seem strained, artificial and immoral to classify children with cars, houses and machinery" (p. 172), but this is quite immaterial to the descriptive success of his theory.

Towards the end of the book Becker uses the economic approach to give an account of social interactions generally, and of family relations in particular. He then employs the most striking result of this application of the new theory of consumer demand to solve what he takes to be the central empirical problem of sociobiology. Becker's strategy for incorporating social interaction as a determinant of individual behavior is simply to add terms representing the characteristics of others that affect the agent's output of household goods. The production function for these goods is

$$Z_j = f_j^i(x_j, t_j, E^i, R_j^1, \ldots, R_j^r),$$

where "$R_j^1, \ldots, R_j^r$ are characteristics of other persons that affect [the agent, i's] output of commodities. For example, if Z_j measures i's distinction in his occupation, $R_j^1 \ldots, R_j^r$ could be the opinions of i held by other persons in the same occupation" (p. 256). Of course, these characteristics of others can be changed by i's own efforts, and these effects can be formalized in a "production function . . . that depends partly on the efforts of i, and partly on other variables." This model is then employed to produce the determinants of intrafamily relations, charitable behavior, multiperson interactions, envy, hatred, and so on. Most important among Becker's results is a theorem deduced from these assumptions about the behavior of members of a family. According to this theorem, if the head of the family loves and cares about its members sufficiently to transfer resources to them (for maximizing their own individual utilities), then all the members of the family are equally motivated to maximize family opportunities and to take fully into consideration the external effects for good or ill of each of their individual acts on the well-being of all the other members of the family. In other words, sufficient love by one member of a family, as manifested in his caring for them and willingness to transfer his productive output to them, leads all members by an invisible hand to act as if they, too, loved everyone. Of course, this theorem is stated more formally and proved in the text, but the intuitive foundation of the proof is easy enough to see. If the head loved every member equally, then he would vary the amount of resources transferred to a selfish member of the family who increased his utility at the expense of another by reducing the level of transfer to the selfish member and increasing it to the exploited member. For obvious reasons, Becker quaintly labels this result the Rotten Kid theorem. Becker employs this result crucially in the final chapter of his work, in which he sketches out an account of the connection that sociobiology can be expected to make between the social sciences and the life sciences. For the purposes of our general assessment of Becker's economic approach it will be worthwhile to lay out this treatment in some detail.

Like a few other economists, Becker has noticed that "the approach of the sociobiologists is highly congenial to economists since they rely on competition, the allocation of limited resources—of, say, food and energy—efficient adaption to the environment and other concepts used by economists." Yet, he points out, "sociobiologists have stopped short of developing models having rational actors who maximize utility functions subject to limited resources. Instead, they have relied solely on the 'rationality' related to genetic selection: the physical and social environment discourages ill-suited behavior and encourages better-suited behavior. Economists, on the other hand, have relied solely on individual rationality, and have not incorporated the effects of natural selection" (p. 284). Becker believes that these two sorts of "rationality"—natural selection and utility maximization—can be brought together and that, since utility maximization exhausts the determinants of human behavior, a sociobiological account of the adaptiveness of utility maximization will *ipso facto* explain all of the consequences of such maximization, i.e., all human behavior. Of course, Becker is cautious in his expression of so grand a claim: "If natural selection and rational behavior reinforce each other in producing speedier and more efficient responses to changes in the environment, perhaps that common preference function [which is the cornerstone of Becker's new approach] has evolved over time by natural selection and rational choice as that preference function best adapted to human society. That is, in the short run the preference function is fixed and households attempt to maximize the objective function [the production function] subject to their resources and technology constraints. But in the very long run, perhaps those preferences survive which are most suited to satisfaction given the broad technological constraints of human society (e.g., physical size, mental ability, etc.)" (p. 145). If this claim about the relation between the determinants of human behavior and the forces of selection can be made out, it will have at least one very general and one very special ramification for Becker's theory. Generally, it will clothe the theory in the mantle of respectability that any theory subsumable under a well-established broader theory acquires. Specifically, this sort of connection provides a further justification for Becker's key claim that all human beings share the same fundamental tastes and that differences between their actions only reflect differences in their means of satisfying them, for the commonality of our fundamental preferences will be as expectable in the face of the forces of evolution as the commonality of our fundamental biological structure. Moreover, this alliance will also have an important advantage for sociobiology. The sociobiologist is often accused of being a biological determinist, of providing explanations for human behavior that exclude appeal to belief or choice, desires, and decision as the determinants of any part of the behavior he purports to explain. If a theory that explains all human behavior in terms of these variables can be shown to be compatible with, and perhaps even to follow from, sociobiological theory, then this sort of objection will have been completely undercut. Moreover, if this demonstration of compatibility can be made out by solving the chief theoretical problem of sociobiology, then the strength of the hypothesis that these two theories can be synthesized is greatly enhanced. And this is exactly what Becker attempts to do.

Becker proposes to solve what he calls "the central problem of sociobiology" by employment of the economic theory of social interactions briefly described above. This "central problem" is that of explaining the biological selection of altruistic behavior, and Becker is right to so describe it. The problem of accounting for individual behavior which tends to increase the fitness of *others* is patently a serious one for any theory which purports to explain all behavior as the consequence of forces maximizing the individual's chances of survival and fitness. By and large, sociobiologists have attempted to bring their theory to terms with the phenomenon of altruism through the construction of a theory of kin selection, which reveals altruistic acts to have adaptive consequences for the close genetic kin of the altruistic agent. This theory is the subject of detailed discussion in Chapter 8 below. But Becker alleges to solve this problem by showing "that models of group selection are unnecessary since altruistic behavior can be selected as a consequence of individual rationality" (p. 284). All Becker needs to show this possibility is the Rotten Kid theorem of his theory of social interaction. According to this theorem, in a family setting, or in any other group in which resources are transferred from a head or parent to other members or children in a way that always maximizes the group's total income, there is no adaptive advantage in the egoistical behavior of any of the recipients, for this will only reduce transfers from the head to him and increase them up to the point of compensation for his victim. Moreover, the Rotten Kid theorem also implies that there can be circumstances in which the altruism of the head may increase his own fitness more than would egoism on his part, for the beneficiaries of his altruism are discouraged from harming him, lest they forfeit the source of their transfers. Since under such circumstances, altruism may increase personal fitness because of its effect on the behavior of others, there is no formal incompatibility between its existence and the claims of sociobiology. Moreover, since the transfer of resources that constitutes altruism is the sort of action which an agent who heads a "household" unit is motivated to perform by his desire to maximize his own utility, it follows that behavior can be explained both as the natural result of selective forces and as the outcome of deliberation and choice. Of course, even though Becker's treatment is far more detailed and formalized than this description has suggested, this solution is far from entirely disposing of the problem of altruism, especially among "lower" species than *Homo sapiens*. Nevertheless, it does enhance Becker's claim that "both economics and sociobiology would gain from combining the analytic techniques of economists with the techniques in population genetics, entomology, and other biological foundations of sociobiology. The preferences taken as given by economists and vaguely attributed to 'human nature' or something similar—the emphasis on self-interest, altruism towards kin, social distinction, and other enduring aspects of preference—may be largely explained by the selection over time of traits having greater survival value. However, survival value is in turn partly a result of utility maximization in different social and physical environments" (p. 294).

Unlike many proponents of the explicit use of maximizing hypotheses in the explanation of all human behavior, Becker cannot be accused of failing to sketch out at least some of its details. He has not only shown that his hypothesis will formally account for all the data hitherto available that might bear on his claims, but he has even set forth previously unnoticed tests of his version of the theory which would seem further to confirm it. Moreover, the version he advances has clear advantages over conventional economic theories of human behavior because it circumvents the problem of specifying preferences and requires the attribution of a single unchanging structure of wants and needs to every human being. It is a structure which, in light of our knowledge of the biological requisites of life, seems quite reasonably attributed to all persons. It eliminates the requirement to identify wants and needs for each case in which we wish to apply a maximization hypothesis. Moreover, his appeal to variations in beliefs about the productive efficiencies of alternative means of satisfying these wants and needs seems to be one with sufficient richness and scope that we may be confident in our abilities to identify them even in the absence of a full fledged theory of human cognition. Finally, Becker's variation on the traditional theory enables it prima facie to explain a wider range of economic as well as noneconomic actions than the traditional one.

The question remains, however, of whether this theory has any *real* (as opposed to merely prima facie) explanatory power. After all, it is only a matter of logical ingenuity, and no real scientific achievement, to erect an axiomatic system that is compatible with any finitely large body of singular causal claims. Indeed, the problem is that too many such systems are easily constructible, and they all have the same prima facie explanatory power; for they may all have the same body of data as their deductive consequences. What features does Becker's theory have that will distinguish it from this vast set of axiomatic systems as one with real, as opposed to merely prima facie, explanatory power? Attempting to answer this question reveals that this best-case economic theory of all human behavior turns out to have the same defects as the theories which it was erected to supersede.

At first blush one might suppose that credibility is lent Becker's theory by the fact that it constitutes a set of general statements relating variables that we know independently of the theory to determine human action, and that it enables us to express these general statements and their interconnections in the undoubtedly scientific terms of differential calculus. Of course, conventional theory may also be expressed in the language of differential calculus, and this is undoubtedly one of the sources of its prestige among the social sciences. But it is also one of the sources of its hold on our imagination, in spite of its practical sterility. The same, it turns out, holds true for Becker's theory as well, despite its apparent advantages in traditional problems circumvented or solved. Why are these theories susceptible of expression in the language of calculus?

The economic approach employs an explanatory strategy which is certainly the best established in the whole of science, for this strategy characterizes all the best scientific theories since the time of Newton. Like Newtonian mechanics—or the synthetic theory of natural selection, for that matter—the economic approach treats the objects in its domain as behaving in such a way as to maximize and/or minimize the values of certain variables. This strategy is especially apparent in Newtonian mechanics when that theory is expressed in so-called "extremal" principles, according to which the development of a system always minimizes or maximizes variables that reflect the physically possible configurations of the system. In the theory of natural selection this strategy is exemplified in the assumption that the environment acts so as to maximize the rate of proportional increase of the fittest hereditarily similar subset of a species. This strategy is crucial to the success of these theories because of the way it directs and shapes the research and applications that are motivated by the theories. Thus, if we know that a system always acts to maximize the value of a mechanical variable, for example, and our measurements of the value of that variable in an experimental or observational setting diverge from the predictions of the theory and the initial conditions, we never infer that the system is failing to maximize the value of the variable in question, but assume that our specification of the constraints under which it is actually operating is incomplete. In the case of mechanics, attempting to complete this incomplete specification resulted in the discovery of new planets, for example, and eventually in the discovery of new laws, like those of statistical thermodynamics. Similarly in biology, the assumption of maximizing fitness leads to the discovery of forces previously assumed to have no effect on genetic variations within a population, and more important, led to the discovery of genetic laws that explain the persistence in a population of such apparently maladaptive traits as sickle-cell anemia. It is because these theories are "extremal" ones that differential calculus may be employed to express and interrelate their leading ideas; and it is because the economic approach is an avowedly extremal theory, asserting that the systems it describes maximize utility or some surrogate, that it can be couched in the language of differential calculus. (It is the extremal character of the theory, and not the fact that it deals with"quantifiable" variables like money, that makes economics a quantitative science, in appearance at any rate. Becker's improvement on traditional theory reflects this fact just because he is able to extend this formalism beyond the range of observably quantified variables.)

What is more important about such theories than that they all employ differential calculus, is that by virtue of this extremal character, they are all committed *to explain everything in their domains*. Because of the claim that systems in their domains always behave in a way which maximizes or minimizes some quantity, the theory *ipso facto* provides the explanation of all of its subjects' behavior, cites the determinants of all its subjects' states. There is no scope for treating such a theory as only a partial account of the behavior of objects in its domain, or as a description of some of the determinants of its subjects' states; for any behavior that actually fails to maximize or minimize the value of the privileged variable simply refutes the theory *tout court*. Thus, the persistence of a maladaptive hereditary trait (which was genetically independent of an adaptive one) over a *large* population randomly interbreeding with one another would not show that some other forces acted on such population in addition to and besides selective forces; it would be taken to refute the theory of natural selection altogether. And this is because the theory asserts that everything that happens to its subjects results in maximizing the rate of increase of their fittest subspecies. Indeed, it is one of the most common objections to evolutionary theory that a "story" can always be told that explains any possible increase or decrease of a hereditarily linked subspecies by appeal to selective forces *known or unknown*. In other words, the pervasive character of extremal theories is but the other side of the coin from their insulation against falsification. Now, all theories are strictly unfalsifiable, simply because testing them involves the employment of auxiliary hypotheses. But extremal theories are not only insulated against strict falsification; they are also insulated against the sort of falsification that usually leads to modification of theories instead of auxiliary hypotheses. In the case of a nonextremal theory falsification may lead either to revision of the description of test conditions, or to revision of the theory by the addition, for example, of new antecedent conditions to its generalizations, or new qualifications to its *ceteris paribus* clauses. But this is not possible in the case of extremal theories. The assumptions of theories like Newton's, or Darwin's, or neoclassical economics', do not embody even implicit *ceteris paribus* clauses. Economic theory does not, for example, assume that agents maximize utility *ceteris paribus*. With these theories the choice is always between rejecting the auxiliary hypotheses—the description of test conditions—or rejecting the theory altogether. For the only change that can be made in the theory is to deny that its subjects invariably maximize or minimize its chosen variable. This, of course, explains why high-level extremal theories like Newtonian mechanics are left untouched by apparent counterinstances, and are superseded, not by qualified versions, in which antecedent conditions are added, but by utterly new extremal theories, in which the values of very different variables are maximized, or minimized, or by new theories of a nonextremal type.

Extremal theories have become an important methodological strategy because of the success of the earliest of them, Newtonian mechanics, but also because their insulation from falsification has enabled them to function at the core of research programs, turning what otherwise might be anomalies and counterinstances into puzzles, opportunities for extending the domain and deepening the precision of the extremal theory. (I have used Kuhnian language here, and we may speculate that much of what Kuhn finds characteristic in "paradigms" is really a consequence of the fact that our most successful theories, since Aristotle, have been extremal ones.) But a necessary condition for the success of extremal theories and their associated research programs is that there be at least some guide to or agreement on independent specifications, characterizations, or ways of identifying the determinants of their subjects' extremal behavior. These guides are themselves theoretical claims, but it is essential that they be distinct from the theory's claims about its subjects' maximizing or minimizing behavior. Two examples from mechanics and evolution will make this point clear.

Treating the Newtonian force law, $F = ma$, as a synthetic proposition governing the behavior of bodies requires that an independent way of specifying mass, m, be provided; and Hooke's spring law provides such a specification. This independent specification not only preserves the contingent character of the force law, but enables us to emply it to explain the operation of springs, projectiles, fluids, and much else that mechanics accounts for. A well-known criticism of the extremal claims of the theory of evolution is that they are empty tautologies because the theory provides no independent specification for the concept of fitness, the variable whose rate of change is maximized within evolving populations. Since higher levels of fitness are cited by the theory to explain greater reproduction rates among populations, these same rates of reproduction cannot provide the independent specification of different degrees of fitness. Accordingly, unless such an independent specification is available, the theory is guilty as charged of being empirically empty. Although showing how this concept can be independently specified, employing generalizations from other biological theories, is so complicated that it would take us too far afield (but see Chapter 7 below), the crucial point is that this sort of identification is possible, and its accomplishment enables us to extend and deepen the theory in the way that Newtonian mechanics has been over the centuries.

This same requirement—of providing independent specifications for the explanatory variables in its theory—must, as noted above, be imposed on the economic approach as well, lest it turn out to be the empirically empty theory that traditional economics' detractors have always accused its predecessor of being. In fact, the traditional objection that economic theory must appeal to changes of taste to explain away apparent disconfirmations, but that these tastes are exogenous to the theory, is a variant of the claim that the theory is unfalsifiable because of the absence of anything like agreed-upon specifications for its explanatory variables, like taste. Can Becker's theory meet this challenge of providing the required specifications? The answer to this question poses several dilemmas for any theory like Becker's. On the one hand, it appears, the theoretical variables that explain utility maximization can be given such a specification. These variables include the basic wants and needs that are common to all human beings, beliefs about market prices and shadow prices, and information about productive processes. Insofar as we have an ability to specify people's wants and needs, beliefs and knowledge, we have the requisite specifications. Unfortunately, as Becker recognizes, the literal interpretation of his theory that such commonsense specification produces is, like traditional economic theory, highly disconfirmed. People do not literally, consciously, make decisions about actions by the employment of principles which parallel or reflect his formalism. Becker guards himself against this obvious objection by noting that "the economic approach does not assume that decision units are necessarily conscious of their efforts to maximize or can verbalize or otherwise describe in an informative way reasons for the systematic patterns within their behavior" (p. 7). In other words, the ordinary interpretation of the theory which commonsense specification seems to suggest is ruled out. We are to treat all of those psychological and cognitive terms that figure pervasively in this book as theoretical terms which do not commit us to the cognitive theory of common sense. Terms like "preference," "decision," "choice," "risk," "information," to which Becker appeals to explain marriage, fertility, social cooperation, distribution of time between labor-market and other activities, discrimination, criminal and police activities, etc., are not to be understood as involving the ordinary sense we accord them, on pain of the theory's being plainly false.

The first question this defense of the theory raises is one of perhaps only terminological interest: can we still call Becker's theory, or any theory like it, an *economic* theory that explains all human behavior as the product of *choices* among available alternatives in order to satisfy *desires*? The economic approach to human behavior is supposed to be contrasted with other theories that explain behavior in terms, say, of conditioning, operant and classical, or unconscious and subconscious forces, or again in terms of the force of institutions, traditions, roles, and other holistic or societal forces. And the contrast between economic theory and these other approaches is that the former theory takes reasons seriously as the determinants of behavior, instead of as rationalizations or epiphenomena. In disallowing this interpretation of the theory, Becker has saved it from evident falsification, but at the cost of depriving himself of the right to call his result an *economic* approach to human behavior, which explains it as *rational action*. It must be emphasized that unlike some traditional economists, Becker does not eschew the aim of explaining the behavior of individual agents and cannot take comfort in the traditional economists' rationale for neutrality about the psychological theory that underlies their account of individual behavior. His explicit aim is to extend the theory to account for all facets of the actual behavior of individual agents, and not just for the economic aspects of the aggregation of all their behavior. Quite independent of the needs for independent specifications of the explanatory variables of his theory, its neutrality with respect to an underlying cognitive psychology makes Becker's claim to provide an *economic* theory of human behavior hollow at best. Furthermore, it turns out to cast Becker's claims to provide a *theory* of human behavior, of any kind, seriously in doubt.

To see this, let us ask what the intended domain of Becker's theory is. The simple answer—all of human behavior—is not really very helpful. Can Becker literally intend the theory to explain everything that human beings do, from dilating the pupils of their eyes, to discovering quantum mechanics? It seems natural to assume that in describing his domain as all human behavior, Becker (or any social scientist with an all-encompassing theory) means to exclude that portion of behavior common to humans and to non-humans—reflex behavior, digestion, behavior that in some sense or other can be described as involuntary. After all, we have fairly well-developed theories for this range of behavior, and they are plainly within the realm of the biological sciences. Furthermore, it seems equally clear that this behavior is not the result of rational calculation, of choice among constrained alternatives. So presumably a theory of human behavior can be excused from having to explain it, on the grounds that such a theory's domain is behavior common and peculiar to humans (and perhaps signing monkeys). It is therefore something of an embarrassment that Becker not only has no grounds to excuse himself from explaining reflex phenomena, for example, but what is more, his theory can in fact at least formally explain it. That is, it is highly probable that our reflex behavior maximizes or minimizes the value of some variable or other that reflects well-being, subject to constraints of some kind or other; and in the absence of interpretations for the variables of Becker's theory, who is to say that the variables of his theory are not the very ones that do account for reflex behavior? In other words, Becker's theory may not be a theory of *distinctively human* behavior at all. This surprising result is a consequence of the fact that Becker's theory is an extremal one, and also one which provides no interpretation for, no independent specification of, its explanatory variables. As a result, there literally is no part of the behavior of humans for which it *formally* cannot account. But this sort of formal ability to account for anything is just what characterizes that vast set of axiomatic systems that are consistent with all the data but do not really explain any of it.

Presumably, we can restrict the domain of a theory like Becker's by stipulating that it is to explain, not all behavior of humans, but all *human* behavior, all behavior distinctive of humans. But how can we characterize this class of behavior? Ultimately, the only answer to the question of what is the domain of a theory is that its domain is all those events which the theory can explain and predict to agreed-upon levels of accuracy. This answer is unsatisfactory and is saved from circularity only by the fact that the levels of accuracy and the deviations from accuracy can be explained by other, more general theories. Thus, we know that the domain of Newton's theory is all systems moving at speeds by comparison to which light's velocity approaches infinity and all systems of such a size that Planck's constant approaches zero in comparison. And the domain of Mendelian genetics is given by all genes whose locations on chromosomes are within a certain distance from one another as determined by molecular genetical considerations. The domains of these theories are determined by their successors, relativity and quantum theory, and molecular genetics, respectively. Prior to the provision of successor theories the domains of the older theories were roughly mapped out by coordinative definitions and correspondence rules of greater or lesser precision which connected the explanatory variables of the theories to observable phenomena. And as we have noted, the longevity of extremal theories lies in the fact that predictive discrepancies were invariably treated as occasions for changing these correspondence rules. These rules provide the independent specifications of explanatory variables required by any extremal theory that has real explanatory force. By parity of method, ultimately, the domain of a theory of distinctively human behavior like Becker's can be characterized only by another theory or theories to which it is reducible. But pending the appearance of these theories, its explanatory power turns on the provision of rough-and-ready correspondence rules or bridge principles relating its explanatory variables to agreed-upon observations. Becker offers no indication of what these independent specifications for his variables are, and indeed goes on to treat them as uninterpreted ones; moreover, there is good reason to think that no such specification can be forthcoming in any case, just because of the extremal character of the theory itself. And this is why it is not only not an economic theory, but not even a theory at all.

Since, as an extremal theory, it purports to explain all human behavior, it must obviously account for buying and selling, working and leisure, marriage and divorce, voting and sharing, as well as for verbal behavior—speaking, writing, and all other acts of communication—of agreement or dissent. Since everything that humans do—everything that is, let us assume, distinctively human—they do in order to maximize utility, it follows that any part of their observable behavior that we hit upon for characterizing (albeit indirectly) the explanatory variables of the theory (agents' beliefs, their common preferences, their level of information about productive processes and other environmental facts) will itself reflect the operation of every one of these variables. Why? Because every one of them determines the level of utility maximization, and all agents' behavior always maximizes. If the theory is correct, we cannot separate these variables, either in principle or in practice, from one another just because the theory accounts for all human behavior. Notice that if the theory did not purport from the outset to account for all behavior, then we might be able to treat the range of behavior for which it did not account as a source of behavioral characterizations of some or all of its explanatory variables. And furthermore, if the theory were not an extremal one, requiring that behavior always maximize utility, then we could use our diagnoses of cases in which behavior failed to do so as indirect but independent guides to the specification of the explanatory variables. Since a theory like Becker's is an extremal theory about all human behavior, it follows that neither of these two sources of independent characterization is open for the specification of its explanatory variables. And if, as I have suggested, it is in the nature of extremal theories to be committed to explaining all of the behavior of systems in their domains, then short of narrowing the theory to only a portion of human behavior—say economic action, for instance—or surrendering the claims about inevitable maximization of utility, the economic approach to human behavior cannot satisfy a necessary condition for being a theory with real explanatory power: the absence of the required independent characterizations seals it hermetically from both falsification and application.

The upshot is that as it stands, Becker's account may be neither an *economic* theory, since it does not explain behavior in terms of rational choice properly understood, nor a theory of distinctively *human* behavior, since its lack of interpretation makes it formally applicable to behavior which is not peculiarly human. Nor, in the end, may it turn out to be a *theory*, as opposed to one of an indefinitely large number of axiomatic formalisms whose lack of interpretation makes them compatible with any occurrence at all, and thereby deprives them of any real explanatory power at all.

In a way, Becker recognizes these problems, for describing the advantages of his approach, he writes that it does not

> take refuge in assertions about irrationality, contentment with wealth already acquired, or convenient ad hoc shifts in values (i.e., preferences). Rather it postulates the existence of costs, monetary or psychic, . . . costs that may not be seen by outside observers. Of course, postulating the existence of costs closes or "completes" the economic approach in the same, almost tautological, way that postulating the existence of (sometimes unobserved) uses of energy completes the energy system, and preserves the law of the conservation of energy. Systems of analysis in chemistry, genetics, and other fields are completed in a related manner. The critical question is whether a system is completed in a useful way: the important theorems derived from the economic approach indicate that it has been completed in a way that yields much more than a bundle of empty tautologies in good part because . . . the assumption of stable preferences provides a foundation for predicting the responses to various changes. (p. 7)

Of course, Becker is right to recognize that the derivation of nontautologous theorems is a necessary condition for the usefulness of a theory, but it is plainly not sufficient, for we may construct any number of systems that have the same finite set of nontautologous theorems. Moreover, the appeal to stable preferences, which represents one of the chief virtues of his theory, is not enough to distinguish it from the class of nontheories of human behavior. Becker believes that the account of the character and components of these preferences is to be given by "the sociologist, psychologist, and probably most successfully by the sociobiologist" (p. 14). But if the sociologist or psychologist deals with behavior, already captured for Becker's theory, in his account of the stable preferences, the result will not provide the independent specification of explanatory variables that Becker needs, and in any case will not confirm the claim that human behavior is the product of rational decision, as opposed to the working out of, say, noncognitive drives responding to various schedules of reinforcement. Do the sociobiologist and the biologist hold out a better hope for Becker's theory? The alliance he seeks to forge between the economic approach and evolutionary theory strongly suggests that he thinks so. And it seems plausible to suppose that biological theory, and not psychological or sociological theory, can provide an enumeration of the stable preferences, the needs and wants, common to all humans that Becker postulates. It can do so in the provision of a list of necessary conditions for evolutionary fitness of individuals; but since the environments in which individual *Homo sapiens* live are so different across space and time, this list will be highly nonspecific and general to the point of banality if it is to be truly and universally attributable to all agents. Moreover, insofar as evolutionary theory already stipulates that individual organisms respond to environmental conditions in ways that maximize their inclusive fitness, that theory already preempts everything that Becker claims for his theory. That is, everything we do can already be explained by a simple-minded application of the theory of natural selection (because it is an extremal theory, of course). Yet we do not consider that theory an explanation of all of human behavior. What reason does Becker offer us, consistent with his own neutrality about the nature of the determinants of human behavior, to suppose that his theory is any more of an explanation of human behavior? In the end the appeal to sociobiology will do no good, for we want, and Becker wants, more from a theory of human behavior than the claim that we all try to survive.

Explanations of human behavior invariably begin with the assumption that we are purposive creatures, and not just purposive but conscious, intending, deliberating agents. Empiricism demands that if we cite purposes and means to their attainment in the explanation of action, these items must be nomologically related. Traditional economic theory represents the most systematic attempt to apply the empiricist imperative to our explanatory assumptions. The teleological character of human action is what enables the economist to state his claims in terms of differential calculus, and thereby clothe his theory in the garb of especially unquestioned respectability. But the intentional character of human action and its determinants deprives the theory of our confidence, for we recognize that on the commonsense interpretation of the causal variables, they simply do not interact with the regularity or precision that the theory teaches; attempts on the part of economists and others to provide specifications for these variables different from our commonsense interpretations and yet independent of the very hypothesis in which they figure have failed equally to sustain our confidence in the truth of maximizing hypotheses. In the face of this problem economists have adopted the course of simply refusing altogether to give an interpretation to their explanatory variables. Thus, one pair of contemporary commentators writes: "A model involving preferences and information, for instance, will ordinarily offer no explication of what preferences and information are, beyond what it says about their structure. The structure itself must be specified with the precision needed for mathematical reasoning."[13] Becker's theory provides a clear example of this, even as he condemns the dependence of older theories on exogenous "tastes" as empty and illicit. In effect, by suppressing all interpretation for their variables while retaining their structural relations as components of a maximization model, economists have chosen to surrender the pretension to explain human behavior as the consequence of intentional variables while retaining the teleological features of the explanation, features which it shares with, for example, the botanical theory of heliotropism. This is what gives the uninterpreted formalism of their theory the flavor of a systematic theory of purposive phenomena, one which we continue to treat as a theory of human action instead of as merely a teleological theory of human behavior, for much the same mistaken reasons that people treat the admittedly teleological theory of natural selection as a theory attributing design or intentional purposes to nature. While misunderstandings in evolutionary theory reflect the *false* belief that nature is in some sense intentional, treating the nonintentional but teleological theory of economic behavior as one which accords intentionality to the causes of behavior is a mistake which rests on the *true* belief that human behavior is intentional.

Once we recognize that, shorn of its interpretation as an account of the effects of beliefs and desires, the theory is no longer a theory of human action, and that saddled with this interpretation the theory seems undoubtedly false, we are faced with a serious problem. For we must surrender one or more of a number of well-entrenched commitments. We can surrender the idea that human action is any sort of maximizing behavior, but to do that seems tantamount to surrendering the view that it is purposive or teleological (in at least the nonintentional sense). We can surrender the notion that distinctively human behavior is determined by or solely by beliefs and desires, ordinarily understood. This is in effect what the economist unintentionally does when he refuses to interpret his formalism. But to do this is to surrender the common assumption that at least some behavior is action, and what is more, it undercuts that vast body of singular causal claims about actions and reasons that we feel certain about, and psychologically could never surrender. Finally, we may wish to give up the search for laws connecting reasons and actions of the sort empiricism seems to demand we provide. But to do this is not only to reject empiricism but to leave open the questions of whether and how the citation of reasons really does explain the actions with which they are connected, and why human beings are not subject to the same principles of nomological determination to which everything else in the universe is subject. This set of apparently inconsistent convictions, no one of which seems expendable, generates the empiricists' chief problem of explaining in a plausible way why the social sciences are so much less well developed than the natural ones.

5

Actions, Reasons, and Natural Kinds

The failure of economics, psychology, or any other science to produce nomological generalizations that will underwrite the singular causal judgments of common sense or social science demands explanation. Some explanations—like Durkheim's, for example—can be ruled out, it seems, on the ground that they undercut statements the truth of which we would preserve against any potential explanation of the failures of social science that was incompatible with them. Equally unacceptable to the empiricist would be explanations which trade on the falsity of empiricism and on the assumption that there are other standards of knowledge besides those reflected in natural science and that social science satisfies these standards. Traditionally, empiricists have sought to slip between these unacceptable alternatives by appeal to the complexity of human behavior and the constraints on our opportunities to acquire controlled and systematic information about it. Since the time of Mill empiricist philosophers have devoted themselves to demonstrating that there are no logical or conceptual obstacles to the existence of laws of human action, but have quite consistently left the actual provision of such laws to practicing scientists. The fact that no such laws have been provided in over a century of empiricist-inspired investigation by social and behavioral scientists can of course be passed off by appeal to these practical difficulties and to the relative youth of the empiricist enterprise in the study of human behavior. After all, a hundred years is not really a very long time on the scale on which the rate of substantial scientific discoveries can be plotted. If no laws connecting reasons to their causal consequences in action have yet been provided, this is only a reflection of the great practical difficulties and is likely to be overcome in the long run, over a period whose length may not be expected to be much shorter than those which have characterized the great revolutions of physics, or chemistry. After all, humans are much more complicated than anything else studied by the methods of natural science, and we cannot expect more rapid results in this case than in the study of simpler systems.

But while empiricists may console themselves with this perspective, it has failed to convince their opponents, who see in the absence of laws of human action a reflection of their logical impossibility. More important, the explanation of the failings of social science on the basis of the complexity and intractability of its subject matter betrays important disanalogies with the explanation of the early failures and slow progress of nascent physical science. The most important of these disanalogies bears on differences between our knowledge of the truth of a host of singular causal statements about human action and the early natural scientist's relative ignorance of such statements about the causal determinants of the events he set out to explain. Throughout the history of natural science we have been faced with phenomena that needed explanation, and although such explanations ultimately involved redescription of the events to be explained and their subsumption under laws, the first task of the natural scientist invariably involved forming hypotheses about their causes and framing rough generalizations relating these hypothesized causes to their effects. He then attempted to sharpen these generalizations, filling out *ceteris paribus* clauses, unifying and synthesizing the generalizations with those explaining other related phenomena, and often reaching the point of redescribing the explananda phenomena and the explanans phenomena in terms utterly foreign to their ordinary descriptions. The slowness of scientific progress is a reflection of the fact that very often the hypotheses about what the causal determinants were for a given phenomena were *wrong*. But such conclusions were reached only after a long, sustained attempt to frame the required generalizations. The failure, after century-long attempts, to frame exceptionless generalizations that would simply and precisely account for the events to be explained in terms of the causal variables then accepted eventuated in what are now fashionably described as paradigm shifts, in which new causal variables were hypothesized and the entire process began again.

The disanalogy between the history of science as thus sketched and the history of social science is that in the latter case we believe that we have been in possession of the relevant variables literally since time immemorial. The conviction that desires and beliefs cause actions was already ancient when Plato expressed it in the *Phaedo*. If just *one* of our everyday judgements about the determinants of action in reasons is true, then we are already in a vastly superior position to that of any natural scientist at the outset of his research. For we already know what the causal variables are. Never mind how we know this, whether by introspection or some other process not available to the natural scientist, just so long as we know it, we must trace back the history of failure to provide the relevant generalization not just to the first self-conscious attempts of the marginalist economists, but to the first instant that the unshakeable conviction appeared that something happened because of someone's reasons. If we were to place natural and social science on a scale permitting reasonable comparison of their rates of progress, their rates would be equal if the technical concepts of Newtonian mechanics (and not just their commonsense reflections) had been known throughout the world to every human being of normal intelligence since the dawn of man and yet still had not eventuated in any of Newton's three laws. Is human behavior that much more complicated than the behavior of natural systems? Of course, the affirmative answer is not logically absurd, but it is equally obvious that this answer is unlikely to figure in an explanation of the failure of social science as reflecting merely temporary empirical recalcitrance that will satisfy someone not already wedded to empiricism. When we add the fact that empirical methods have enabled us to frame powerful general laws and theories about matters vastly more complex than the mechanics of medium-sized objects, and vastly more recalcitrant to observation than human behavior, the implausibility of this explanation becomes proportionately even greater. These successes have convinced us that we have in hand the methods of discovering and confirming powerful laws and theories about all manner of complex and observationally inaccessible phenomena. Thus, the application of such methods over the period of a hundred years and more to a body of phenomena in which we believe we can already identify the important causal variables and their effects makes the empiricist's traditional explanation for the absence of laws of human action little short of totally incredible. Furthermore, if the practical impediments to the establishment of a science of human behavior replete with laws and theories of the requisite power are of these dimensions, then they might as well be logical or conceptual impediments for all their practical force. If the complexities of the subject matter are as great as empiricism requires in order to substantiate its explanation of the failures of social science, then it seems unlikely that the devotion of any finite amount of energy over the reasonably foreseeable future is likely to provide the science it assures us is possible. But the more remote and academic the possibility of a science of human behavior, the less and less force can we accord empiricist prescriptions in the actual practice of social research, for we cannot justify the prescriptions by their probable foreseeable results.

In short, the empiricist explanation of the failures of social science to attain the standing of natural science is either hopelessly implausible or hopelessly pessimistic. For the factors of complexity and observational recalcitrance that it trades on are either too weak to account for the failure to be explained or too strong to sustain the expectation that more industry and a bit of genius will eventually retrieve the failures thus explained. This dilemma makes the appeal of anti-empiricist philosophies of social science more and more understandable if not acceptable and, when added to the constraints of a theory of human behavior enumerated at the end of Chapter 4, makes empiricism seem untenable as a philosophy of social science. To be saved from untenability, empiricism requires either a law of human action or another explanation of its absence besides the complexities and recalcitrance of human behavior.

In this chapter I shall begin to detail an explanation consistent with empiricism for the failing of social science that does not trade on appeals to complexity and recalcitrance. The explanation also has the advantage that it makes sense of the history of the last hundred years' search for a law of human action—makes sense of it both by showing that the search was not conceptually misconceived or logically fated to failure, and by showing precisely why the search failed. It will also suggest (in the next two chapters) the direction in which, consistent with empiricism, subsequent work in social science must proceed.

Although economists and other social scientists may be accused of over-hasty generalization in their attempts to formulate laws that will systematize all distinctive human behavior by appeal to maximizing principles, philosophers too have attempted to formulate and assess principles of similar generality. The philosopher's aim in the formulation of such principles has been quite different from that of the economist or psychologist. Philosophers have attempted to uncover the expression of a general relation between reasons and the actions they determine in order to expound and assess claims about rights and responsibilities that are accorded to individuals in the light of their powers of action; to make sense of our attributions of praise and blame, freedom and constraint, our distribution of desert and punishment; to make sense of the commonsense singular explanatory claims we make every day about our actions and the actions of others, their consequents and antecedents. In this last connection, quite independent of the empiricist's claim that these singular claims are causal and therefore presuppose laws, philosophers have argued about whether our ordinary explanatory practices commit us to a synthetic general principle or lawlike statement in any case. The camps who dispute whether such a principle is reflected in our true commonsense claims about particular reasons and specific actions have produced many candidates for such a principle, only to see each claimed either to be false or tautological, just as the principles offered by economists seem to have been. Among the principles offered by philosophers as lawlike claims presupposed by ordinary explanations of human action, however, one stands out as more plausibly true and arguably less trivial than the others.

This candidate for a law of human action is generated by Paul Churchland in the course of an argument to show that "there are some fairly sophisticated nomic principles or 'laws' *specifically* presupposed by our ordinary action-explanations."[1] Now, if the singular statements that constitute our ordinary action-explanations are true, and if they presuppose specific laws that Churchland has a method of isolating, then the empiricist's laws of human action must be available after all. "That our explanatory practices with respect to human actions are presumptive of what is *prima facie* a general law," Churchland writes,

> can fairly easily be brought to light by a systematic examination of those practices. An adequate theory of the logical character of action-explanations must be able to account for the undoubted propriety of the various types of everyday *objections* to which they can be subject, and this fact provides us with a strategy for winnowing out the underlying law, if it happens there is one. We need only examine and classify the types of objections which can legitimately be raised against an ordinary explanatory statement of the form '*x* *A*-ed because he wanted *φ*' in order to bring out the entire set of necessary conditions [i.e., the set of conditions jointly causally sufficient] for the correctness of that explanatory statement (p. 215).

Churchland's strategy for constructing the lawlike statement which he claims is implied by our explanatory practices is particularly attractive because it trades on just those considerations which philosophers have appealed to in order to falsify putative laws, the sorts of countervailing forces which make for exceptions to the truth of these candidates. If Churchland can collect *all* these counterexemplifying considerations and impound them within an exception-excluding clause of a principle connecting reasons and actions, he will have uncovered the lawlike statement presupposed by our common-sense singular claims. In this sense, Churchland's strategy will work, if any strategy will. But of course this does not guarantee that any strategy for detecting and establishing the required general statement will work; and in particular, if there are an indefinite number of different ways in which exception can be taken to the putative law, it will never be open to expression in a finitely long statement, for it will require an indefinitely large number of excluding clauses. And a law that cannot be stated is, presumably, no law at all for the purposes of science. Churchland, however, believes that the number of such exceptions is manageably small.

Churchland begins with the sort of general statement that one might initially generate to sustain the claim that "*x* did *A* because he wanted *φ*" (p. 216):

(x) (ϕ) (A) (if (1) *x* wants *φ*, and (2) *x* believes that *A*-ing is under the circumstances a means for him to achieve *φ*, or contribute to his achievement of *φ*, then *x* does *A*).

It is the content of this general statement that the economist seeks to capture in a quantitative and highly general way in his hypothesis of maximization; and it is the obvious counterexamples to its truth that also drive the economist away from the literal interpretation of his hypothesis as a recognizable variant of this one. The first of these objections turns on the fact that sometimes an agent satisfies clauses 1 and 2 in our general statement without satisfying its consequent, because the agent believes that there is some other action besides A-ing which is a means to bring about ϕ, under the circumstances, and that this action is preferable to A-ing. We may insulate our general statement against this sort of case by adding a clause of the following form to its antecedent:

(3) there is no other action believed by x to be a means for him to bring about ϕ, under the circumstances, which x judges to be as preferable to him as, or more preferable to him than, A-ing.

Notice that the assumption that the agent is correct in his beliefs of this kind is tantamount to the economist's traditional boundary condition of perfect information about the availability of alternatives among which the agent is to choose the most efficient for attaining his preferred goal or maximizing his utility. Of course when clause 3 is added to the generalization, there is no presumption that the agent's belief is correct.

Of course, an agent may satisfy all of clauses 1, 2, and 3, and still not do action A. He may, for instance, have another want besides ϕ which is stronger than ϕ and overrides it. It is to avoid this sort of problem that the economist appeals to the desire to maximize utility, for this desire is supposed to underlie or to generate all other more particular wants for particular ends. To avoid this sort of falsifying circumstance we need to add another clause to our principle:

(4) x has no other wants which under the circumstances override his want ϕ.

By now the strategy has become clear. If we can add a clause excluding each of the kinds of exceptions to the lawlike connection described above, we will have the general statement which underlies our singular causal judgments about reasons and actions. According to Churchland, only two more clauses are required. One excludes cases in which the agent does not perform the action because he is physically incapable of performing the action in question. The other circumvents objections based on the supposition that the agent does not know or have true beliefs about how to do the action in question.

If this enumeration of possible objections showing the falsity of our initial general statement is complete, then, according to Churchland, "the following conditional can plausibly be seen as nomological in character" (p. 221):

L $(x)(\phi)(A)$ (if (1) x wants ϕ,

                       (2) x believes that A-ing is a way for him to bring about ϕ under those circumstances,

                       (3) there is no action believed by x to be a way for him to bring about ϕ, under the circumstances, which x judges to be as preferable to him as, or more preferable to him than, A-ing,

                       (4) x has no other wants which, under the circumstances, override his want ϕ,

                       (5) x knows how to A,

                       (6) x is able to A,

      then   (7) x does A.)

L is generated from our common beliefs about actions and their explanations, as revealed in the kinds of criticisms we might offer of a statement of the form "x did A because he wanted ϕ." L's strength and plausibility derive from the completeness of Churchland's survey of these kinds of criticisms, and from the fact that an overlooked sort of objection can easily be accommodated by adding still another clause to the six antecedents of L. L has the further virtue, not noticed by Churchland, that aside from reflecting common sense it also provides a full specification of the sort of generalizations that underlie the social scientist's systematic explanations of actions in terms of reasons. For when he appeals to a principle like the maximization hypothesis, he tacitly assumes that each of the six conditions in the antecedent is satisfied by the explanandum phenomenon and interprets his maximization hypothesis in such a way as to exclude the circumstances that L excludes explicitly.

Churchland offers arguments to show that L is plausibly construed as a law, as being "nomological in character." He does so by attempting to rebut charges that L is a disguised tautology, or an analytic proposition true by virtue of the meaning of its terms, and by attempting to undercut arguments to the effect that L is false because there remain as yet undescribed but nevertheless undoubted counterexamples to it in the form of the ancient Aristotelian problem of *Akrasia*, weakness of will. Insofar as we countenance the existence or the causal possibility of such cases in which agents know what the best thing to do is, but in the absence of impediments fail to do it anyway, L must be false. Churchland has several rejoinders regarding the bearing of this ancient philosophical puzzle on our concerns. But for our purposes suffice it to say that the hypothesis of *Akrasia* is simply the direct denial that human action, as opposed to mere behavior, is causally determined by beliefs and desires exclusively. As such, the objection merely begs the question at issue unless compelling reasons are provided beyond mere intuition for the possibility it envisions. In any case, I shall assume that Churchland can successfully defend L against both the charges of analyticity and falsity. For my aim here is to show that in spite of the fact that L is *ex hypothesi*, a true synthetic general statement, it is not a law of human behavior, and that the explanation of why it is not one plausibly explains, in a way consistent with empiricism, the absence of any laws of the sort on which social science seeks to ground its true singular judgments.

Although I shall assume that Churchland is correct in denying the analyticity of L, it is important for our purposes to examine how he answers this challenge. He writes:

> The objection that L is flatly analytic. . . [and therefore not an *empirical* general law] may appear to be a serious one given the confessed difficulty of conceiving of a falsifying instance. The objection is highly welcome, however, for its apparent seriousness is readily undermined: on the view being proposed, the appeal of the objection is quite understandable. After all the suggestion is that L is a *deeply entrenched theoretical nomological* central to our understanding of human behavior, and of such states as wanting, believing, and preferring—a basic principle of the conceptual framework in terms of which we conceive ourselves. It is difficult, perhaps impossible to deny L without undermining the conceptual machinery which makes such understanding possible or, better, *constitutes* it, but none of this entails that L is 'analytic' in any sense inconsistent with its being nomological in character. One could not deny the principle of mass-energy conservation without threatening similar havoc in the conceptual framework of modern physical theory. . . .
>
> . . . We can concede that the rejection of L would entail serious conceptual readjustments, but conceptual change is characteristic of theoretical change, and the status claimed for L is that of a *theoretical* nomological. (p. 225)

The cogency of replies of this sort usually turn on denials of a sharp analytic/ synthetic distinction (which Churchland does not himself appeal to), but what is more important than the merits of this passage as an argument against treating L as analytic is its claims about the centrality of L to a theoretical edifice. For if Churchland's claim that L is at least a viable candidate for nomological status is to be substantiated, then this centrality will have to be shown to be either actually or potentially of the same character as that of the "entrenched theoretical nomologicals" of natural science.

As we have noted in previous chapters, it is obvious that something like L plays a central role in our conceptual scheme. Let us make this claim more precise: assuming it is true, L (and the propositions which follow from it upon the substitutions of particular agents, wants, and actions for its bound variables) provides the only systematic means of determining whether each of the seven types of states mentioned in its antecedent and consequent are in fact instantiated on any particular occasion. L's centrality to our conceptual scheme is a reflection of the fact that under their ordinary interpretations its key terms— 'want', 'believe', 'judge', 'prefer', 'knows how', 'is able to', performs an 'action'—are ones for which only functional characterizations are available. That is, their normal, nontechnical meanings are given by appeal to the causal roles which we believe the items denoted by these terms play.[2]

Terms like 'belief' and 'want' or 'desire' seem especially clear examples here. Though there is no explicit noncontroversial definition of these terms, everyone has a rough idea what they mean, and there seem to be clear cases of belief and desire about which no one is seriously in doubt. For example, we ordinarily suppose that nothing could count as a desire unless it led at least on some *possible* occasions to actions that the agent would not otherwise have performed. Here, desire is characterized by one of its effects. Similarly, nothing could count as a belief unless there were at least some *possible* circumstances in which the believer would answer yes to the question whether he had the belief in question. Equally, it is supposed that nothing could count as a belief or desire unless an agent's having it could be brought about in certain ways that involve changing his circumstances. And of course, differing beliefs and desires are distinguished by reference to differing causes and effects which reflect a belief's or a desire's (propositional) content and the degree of its strength. These claims about belief and desire express functional characteristics of these two sorts of mental items. It seems undeniable that our knowledge of the doxastic states and preference structures of other agents is based (inductively or not) on our knowledge of the causal conditions and consequences of these states and conditions. It seems fair, therefore, to infer that these conditions and consequences also figure in specifications of the meaning of claims about desires and beliefs. What nonfunctional characterizations can be offered to supplement such characterizations in our account of the meaning of these terms? A physicalist, of course, might say that among a desire's or a belief's nonfunctional characteristics are the physical properties of the neurological item which constitutes its physical realization. Similarly, an opponent of physicalism might cite their private or privileged position with respect to the agent who has the belief or desire in question. But both of these claims are too controversial to figure in accounts of the ordinary meanings of 'desire' and 'belief.' Moreover, one especially striking feature of arguments to the effect that beliefs and desires are brain states or that they are epistemically privileged in a way incompatible with their being brain states is that both such arguments proceed by appealing to the notions in question under functional characterizations. Beliefs and desires *qua* brain states are to be individualized and identified through the employment of sophisticated scientific instruments whose readings are supposed to be nomologically linked causal consequences of the occurrence of the brain states in question. Similarly, claims to the effect that beliefs and desires are known by their "owners" in a way privileged by contrast with the way others are acquainted with these mental items trades on the assumption that our knowledge of others' mental states is based on criteria of a causal (if not contingent) nature.

In fact, the only noncontroversial conditionals we could offer to characterize explicitly the circumstance under which each of the states described in L are occasioned take the form of the state in question on one side, and one or more (perhaps all) of the other six conditions of L on the other side of the connectives. Thus, for example, the meaning of the concept of want is given by an expression of the following form:

W (x) (ϕ) (A) (x wants ϕ only if clauses (2), (3), (5), (6), and (7) of L obtain.)

In order to turn W into a statement of necessary and sufficient conditions for wanting, instead of a statement of necessary conditions, we must add clause 4 of L. But this clause employs the concept of want itself, and so makes a full characterization of want formally circular. Nevertheless, providing a characterization in terms of conditions only necessary for wants is perfectly legitimate and indeed characteristic of the limitations on the introduction of terms in scientific theories generally: typically it is not possible to provide a noncircular characterization of such terms which is also a sufficient condition of its application, for otherwise they could be eliminated from the theory without loss of explanatory power, and this is almost never the case for theoretically significant terms. A similar characterization for each of the key terms of L is easily constructible, and by virtue of the exceptionlessness of L these characterizations will be the most complete available. This is because the existence of a further necessary condition for someone's having a want, for instance, not mentioned in W would reflect an exception which L would have to be amended to accommodate, in order to remain a true exceptionless general statement.

Definitions like W proceed by characterizing their definiens in terms of the causes and effects of an agent's manifesting the predicate defined. This, of course, is the sense in which the terms of L are all functionally characterized ones. But once the functional nature of the concepts that figure in L is admitted, the centrality of L to our conceptual scheme acquires an explicit and clearly understood sense. Functional characterization presupposes causal connection, which on the empiricist's view, in turn, requires the existence of nomological generalizations (known or unknown). But L provides the sole available candidate for an exceptionless law in which any of the states and conditions in question figure. Accordingly, its falsity would deprive the terms describing these states of the only noncontroversial characterizations available for them.

These considerations both explicate and underwrite Churchland's claim that "the rejection of L would entail serious conceptual readjustments." Not only would our commonsense attributions of the mental states and actions that figure in L be rendered seriously indeterminate, if not meaningless, by the repudiation of L; but without appeal to L, both to explain action and to determine the instantiation of its antecedents and consequent, any social theory that sought to explain action as the effect of desires and beliefs would, so to speak, float on a void. But from these considerations it does not follow that L is a law after all; indeed, they suggest a crucial difference between L and general statements in natural sciences which are accorded this status. Because L not only substantiates the citation of reasons as the *causes* of actions, but also constitutes the sole available criterion for identifying these causes and effects, it fails to be the "deeply entrenched *theoretical* nomological" that Churchland describes it as.

To show this I want to draw a comparison between L and a principle of Mendelian, or population, genetics, and the relations of their respective terms to other general statements in which they do or logically could figure. Now, this analogy between a statement like L and a Mendelian law has been mooted before by physicalist exponents of the identity of the mind with the brain,[3] but they have not discussed this analogy in sufficient detail to show exactly what obstacles face their version of materialism, nor have they employed the comparison to show the difficulties surrounding the entrenchment of L in a scientific theory and the consequences of this difficulty for the pursuit of a scientific theory of human action. We shall see not only that there are disanalogies here which outweigh the analogies, but that the prospects for entrenchment of L in a scientific theory are bleak in the light of these disanalogies between L and Mendel's laws.

Mendel originally offered two laws of population genetics, in contemporary versions of which the gene figures as the cause of differential inheritance, mutation, and replication. For our purposes it may be useful to state the following version of one of these laws:

The Law of Independent Assortment: The presence of a gene for any particular inherited trait does not determine the presence of any other gene.

Mendel appealed to this law to explain why hereditary traits like the height of a pea plant and the texture of its fruit are transmitted across generations in entirely independent proportions. As stated, however, this law turned out to be shot through with exceptions, rather like versions of L that omit one or another of its six antecedent clauses. First it was discovered that in some organisms some genes that code for apparently distinct phenotypic traits were constantly associated across generations. Originally this unexplained "linkage" of genes resulted in the interpolation of a clause in the law of independent assortment which excluded such cases. But it turned out that there was an exception to this exception. Subsequent studies showed that occasionally, linked genes which code for distinct phenotypic traits could be separated from one another in their effects if their locations on a particular chromosome were widely separated, and if this chromosome broke during meiosis. This phenomenon, known as crossover, requires still a further revision of Mendel's law—indeed, some would argue, an abolition of this law in favor of one stated by appeal to a new concept of gene, or to a concept of different genelike units each of which had different functions with respect to heredity and which could function in exceptionless laws of population genetics. Whether the result of separating these units of inheritance, mutation, and replication is a new law, or part of a new theory, that replaces Mendel's, is not important here. What is important is that a "version" of Mendel's law can be stated in which a genelike entity that does not exhibit linkage or crossover can be substituted for the notion of gene as originally understood. This entity is called the "cistron." It is functionally characterized as the genetic unit responsible for inheritance, but unlike the Mendelian gene, it is not also functionally characterized as the unit of mutation and replication. Thus, we may assert the law of independent assortment as an exceptionless regularity about the independence of the cistrons that play a role in the determination of hereditary traits. If our "law" about reasons and their effects in action is in fact an exceptionless general statement, then it is to the law of independent assortment of cistrons, not of Mendelian genes, that L must be compared. Indeed, we can coin a single term which will encapsule all the provisos of the antecedent of L in the way that "cistron" encapsules the exclusions of linkage and crossovers, and so express L as a law relating actions and a single type of (conjunctive) state of the agent which causes them. But although this artificial reformulation of L may render it topographically similar to a law of the assortment of cistrons, there is another important feature of this latter law that mere inscriptional changes cannot bestow on L as well.

The concept of cistron is not the product of discoveries on the molecular level, and was introduced into genetics long before the detailed translation of the genetic code. So it neither raises nor settles questions surrounding the prospects for reducing the theory in which it figures, transmission genetics, to molecular genetics. The appeal to the notion of cistron was founded on purely biological (and not chemical) techniques of breeding and testing for crossing over and linkage. As the notions of linkage and crossing over suggest, however, the notion of the cistron is the product of a progressive localization of the classical Mendelian gene (and its successors) initially to individual chromosomes, and eventually to distinctive regions of particular chromosomes. It was this localization that provided a method *independent* of the law of assortment for determining whether something is a cistron, and whether its behavior confirms the law. Quite independently of advances in molecular genetics, methods of locating cistrons and maps of the location of cistrons on microscopically visible bands of particular chromosomes of particular species were provided: these methods and maps provided both the identity conditions required for the assertion of the existence of cistrons, and the criteria of individuation for them required to provide a test of the general statements in which they figure.[4]

By describing in great brevity the character and methodological situation of the concept of the cistron, I do not mean to suggest that there are no connections between the laws, like that of assortment, in which it figures, and the methods by which we can determine its occurrence. I am merely noting that there is at least more than one general statement available for characterizing cistrons, so that confirming any one of the laws about cistron behavior does not require *immediate* appeal to itself or to one of its direct consequences. As Churchland notes in defense of L, no universal is testable "independently of the entire framework of principles of which it may be an integral part," but at least in the case of the law of assortment, *there are other principles* for localizing cistrons and their resulting maps that in effect provide biconditionals which individuate types of cistrons by their locations on chromosomes. These biconditionals do not provide identities, or decompositions, or structural characterizations, or anything else besides locations for cistrons. It was by the subsequent chemical analysis of the bands of chromosomes at which the cistron is located that their structure, constituents, and the mechanism of their operations that molecular genetics consists in was ultimately provided. It will not be necessary to turn to the relation between the cistron and the DNA out of which it is composed in order to make important comparisons between population genetics at this level of development and our candidate law, L.

The important difference between L and the law of assortment of cistrons is not that we have identities for cistrons in terms of strings of DNA (we do not), nor even that we know what the precise general locations of particular cistrons are. The difference is that we have no independent general characterization of the items mentioned in L that will permit us to determine whether any agent exemplifies one or more of them on a given occasion. This difference is vital to the question of whether a statement like L is or can figure as an "entrenched theoretical nomological," not just because the absence of such characterizations independent of L precludes noncircular tests of L, but more important, because the lack of these characterizations isolates L from generalizations that might causally explain it and from generalizations and singular statements that it might causally explain. The restriction of cistrons to particular bands of chromosomes enables us to explain the law of assortment as a consequence of generalizations governing the process of meiosis (a process characterized in completely nonmolecular and nongenetic terms), and enables us to employ the law of assortment to explain the consequences of meiosis for the inheritance and distribution of observable detectable phenotypes in general, and in particular cases. Our inability to similarly connect L to laws above and below it in a theoretical hierarchy is a direct consequence of the lack of similar independent characterizations for the items that figure in it.

Whereas the localization of genes to chromosomes enables us to connect a general statement describing their independent assortment with independently established generalizations about the multiplication of chromosomes in meiosis and with independently established generalizations about the distributions of observable characteristics of organisms among their descendants, no such entrenchment is possible for L because of the nature of its terms. To see this, notice that in the case of the Mendelian law, theoretical entrenchment is provided by independent specifications of the law's terms, because they enable states of affairs described in its antecedents and consequents to be shown to figure as respectively the consequents and antecedents of other independently established generalizations. Now, for L to be similarly entrenched there must be independently establishable generalizations about, for example, the environmental determinants of belief or desire, and about the consequences, either individual or aggregate, of actions. But, of course, these laws must be independently establishable; that is, it must be at least possible to offer confirmatory evidence which does not presuppose or entail the truth of L. But from this it follows that these laws cannot employ any of the terms in L, such as want, or action, whose meaning, as we have seen, is given by functional characterizations that presuppose the truth of L. On the other hand, if they do not employ such concepts, the fact that L provides the sole means for characterizing its own concepts *ipso facto* excludes the establishment of connections between its terms, and the terms of these generalizations about the determinants of wants (otherwise described, of course), and the consequences of actions (also not described as actions). Thus, if L is truly exceptionless and is the sole candidate for a law about the causal determinants of human action, then the fact that its terms are functionally characterized entails that L cannot be connected with other generalizations in a way that nomological entrenchment requires.

In the face of this conclusion, the most promising line of defense for L's status as a law seems to lie in finding some new characterization of the terms in L which will be independent of L in the way required.

The demand that independent characterizations of the items mentioned in L be provided, together with the assumption that L is in fact an exceptionless truth about the causal determinants of action, places certain significant restrictions on the form that such independent characterizations can take. In particular, it effectively restricts such characterizations to physical, and specifically, neurophysiological ones. If the terms in L are, as I have argued, functional ones, and if L reflects the *sole known* exceptionless general statement relating them to their causal concomitants, then, given present knowledge, the possibility of providing other general *functional* characterizations of any of these same terms is *ipso facto* excluded. Any characterization of the beliefs, desires, and preferences mentioned in L that trades on their propositional content, for example, or on their strength as measured in behavior trades on the implicit or explicit employment of L, and will therefore not

provide a characterization of the required sort: one that is independent of
L. This leaves, as the only source of acceptable characterizations, the physical
structure or the physical concomitants of these items. I say physical structure
or concomitant because the issue here is not (yet) one of reduction or identification that materialism hopes for. It would be enough to provide the required characterizations if we could provide a principle of systematic isomorphism for the states mentioned in L.[5] Such a principle would take the
following form: for each of the types of states mentioned in L, there is a type
of physical state such that for every x and for every time t, x manifests one
of these L-states at t if and only if x manifests the relevant physical state at
t. Reduction is not at issue here any more than it is in connection with the
systematic isomorphism implicit in the localization of (but not identification
of) cistrons to bands of chromosomes. In and by themselves such principles
give us no guide to the structure of items otherwise functionally characterized, though they would provide independent means of determining whether
an agent manifested any one of them. But suppose, as seems reasonable, that
there is no single type of neurophysiological state uniformly associated with
the states mentioned in L. If the number of such states is manageably small,
then the prospects are good that industrious pursuit of the neurophysiological
basis of mental activity will provide at least a manageable disjunction of states
which are jointly isomorphic with L's states. Of course, the larger the number
of distinct types of neurophysiological states which are jointly isomorphic
with L's states, the harder it will be to provide the required characterizations,
and the more impractical will become appeal to such characterizations to connect L to other laws about agents, or to particular instantiations of L.

Here again, it is worth recalling the analogy with Mendelian genetics and
its successor. In effect, the transition from employment of the concept of
Mendelian gene to employment of the concept of cistron in the statement of
the law of assortment reflected just this situation. Through localization it was
discovered that the Mendelian gene, defined as a unit of hereditary transmission, mutation, and replication, is concomitant to three different sorts of
items. It turned out that there was not just one thing which performed the
functions ascribed to the Mendelian gene, but three, and furthermore, that
each of them performed a different one of the three functions hitherto
attributed to the gene. The cistron is the one which performs the function
accorded to genes in the original version of the law of independent assortment, and it was for this reason that it was substituted into the original
law (qualified by the exclusion of linkage and crossover) in place of the gene
classically defined (thus permitting the deletion of the qualifications). In
other words, it was just because *a limited number* of items were found to
correspond to the Mendelian gene that this concept was *replaced* in the
course of the development of genetics. Similarly, I suggest, if neurophysiology disclosed a limited number of types of states disjunctively concomitant

with states like beliefs, desires, and actions (in otherwise normal bodies), then these concepts are likely to be replaced by such neurophysiological ones in the strict versions of laws that govern human behavior. If L is an exceptionless truth, as we have assumed, and if the occurrence of types of states mentioned in its antecedent and consequent and the causal connection between these states is assured by virtue of the systematic isomorphism between these states and a finitely long disjunction of brain states which are causally connected by one or more neurological laws, then L will become theoretically dispensable (just as the heavily qualified law of the assortment of genes became); and it will be dispensed with, just because of the complications that attend the determination of whether its antecedent and consequent are fulfilled by particular agents engaged in particular bodily movements. The upshot is that unless there are relatively simple correspondences, appealing to only a very small number of alternative neurological specifications of the items L mentions, L has no hope of figuring as a "deeply entrenched theoretical nomological."

Moreover, it can be shown that L's pretensions to nomological status are incompatible with the conjunction of (1) the supposition that there is no finitely long list of neurophysiological items associated with each of the terms of L, and (2) the altogether unshakeable belief that at least some of our *singular* causal claims about the relations of particular wants, beliefs, abilities, and their consequences in actions are true. To see this, suppose that none of the relevant correspondences are available, because for any particular type of want, ϕ, for a particular action, A, and for any particular types of beliefs with respect to ϕ and A, no two interpersonal or intrapersonal instantiations of these types of states are correlated with the same type of neurophysiological state. (Anyone who believes that this supposition is too restrictive, and therefore implausible, may increase the number of interpersonal or intrapersonal instantiations of the same type of mental event from two up to any finite number without affecting the argument to be broached.) If we add to this supposition the assumption that one version of materialism is correct, that every *particular* mental state or event is identical with some particular brain state or event, then the supposition can be shown to be consistent with the truth of any singular assertion that some *particular* set of states of belief and desire caused a particular action. If the particular mental states are identical with particular brain states, and the action is identical with a particular bodily movement, then they may be causally connected by virtue of the existence of a neurophysiological law subsuming the physical events which they happen to be identical with, even though it is false that the types of states of belief, desire and action exemplified on this occasion are uniformly exemplified by physical states of this (or some other finite number of) type(s).[6] The upshot of this argument is that particular desires and beliefs may be the causes of actions. Nevertheless, it will not be *by virtue of* their being desires and beliefs of any type that these particular states of an agent cause his actions, nor will the actions be effects *by virtue of* their being actions of any type. Particular mental states of agents and their actions are causally related because they are also types of physical states of the agent, and for no other reason. But this conclusion is tantamount to the claim that there is no constant conjunction of types of desires and beliefs with types of actions, or, in other words, that there is no law relating reasons and actions. Accordingly, on the assumption of materialism, the further supposition that there are no correlations between types of mental states and finitely many disjunctions of types of physical events implies that L is no law.

Nor will the suspension of the assumption of materialism help preserve L. On the supposition that beliefs, desires, and actions are not correlated with a finite disjunction of physical states, L turns out to be either no law at all or else a law that no finite number of scientists could ever have grounds to call a law. If each of the types of states that figures in L is correlated with an infinite number of different types of physical states (provided, of course, that it makes sense to speak of correlation in such a case), then naturally any characterization of these terms of the sort required to entrench L in a scientific theory will be impossible to express, because of their infinite length. What is more, L would itself be deducible from nothing less than an infinite number of laws about the causal relations between these infinite numbers of different types of physical states. On the assumption that a proposition of infinite length which independently characterizes each of the items in L cannot be discovered by a finite number of scientists, it follows that L will never be known to be entrenched in a scientific theory, since independent specification of its terms is not forthcoming. If it is further assumed that a general statement cannot be a law if its truth is known to turn on the truth of either an infinite number of other general propositions, or on propositions of infinite length, then L is no law at all.

Thus, given materialism without expressible isomorphism of brain states and mental states, we may preserve the truth of our singular causal judgments about reasons and actions, but only at the cost of surrendering L as a (knowable) law that underwrites them. On the other hand, without materialism both the possibility of laws connecting reasons and actions and the truth of singular statements they sustain are undercut. The incompatibility of L and the truth of any of the singular causal statements it was constructed to underwrite surely constitutes a *reductio ad absurdum* of the claim that L is a law in the absence of infinitely expressible isomorphism between its terms and neurophysiological ones. But the possibility of such expressible isomorphism beyond the simplest types also leads to the same conclusion: that L has no hope of nomological entrenchment.

L is, in fact, in important ways unlike both the exception-ridden law of the assortment of genes and the law governing the assortment of independently specified cistrons. Because of its exceptionlessness we are not motivated to search for (exceptionless) substitutes, and because of the absence of independent specifications for its terms, we are in no position to test its putative exceptionlessness or to entrench L in a scientific theory. Moreover, the prospects that advances in neurophysiology will provide manageable methods of independent specification are extremely small. Neurological correspondents of states of belief, desire, preference, etc., will have to be several orders of magnitude more complex than the molecular correspondents of the cistron. But we already know that any attempt to stipulate the latter sort of correspondence is almost unmanageably difficult. The degree of redundancy and the number of different functional roles of even the smallest strands of DNA suggest that although we have excellent reason to believe that cistrons are "nothing but DNA," it seems neither likely that we will ever have full statements of the required correspondences nor useful to attempt to provide them.[7] It is sufficient that we know that such correspondences are in principle possible to construct. By comparison, we have no assurance that the parallel correspondences for L's terms are even in principle available; after all, the number of neurophysiological realizations may be infinite. And in any case we have the best of reasons to believe that these correspondences are vastly if not infinitely more complex and unattainable than for the cistron.

Insofar as the possibility of entrenchment as a theoretical nomological statement, whether deep or shallow, is a necessary condition for a proposition's being a law, L is not a law, even if it is, as assumed, a true synthetic general statement. Notice that the requirement for a statement's being a law is not actual entrenchment but possible entrenchment, and not clear centrality to a vast theoretical edifice but only integral connection with one, either at its periphery or at its center or between the two. Our argument has shown that the exceptionlessness of L, coupled with the duty it performs in characterizing the items that it cites, deprives it even of the possibility of such integral connection with other nomological generalizations. But this conclusion raises serious questions. How can a true synthetic general statement fail to be a law? If L is no law, how can the singular causal statements which seem to presuppose it be substantiated as true claims themselves? And finally, how can such a conclusion figure in an argument that preserves the empiricist program in social science, with its demand for generalizations? The answer to all these questions follows from the hypothesis that L is no law because the classes of items that it cites in its antecedent and its consequent are not *natural kinds*.

Consider the concepts of "phlogiston," or "fish," or "race." These notions do not designate natural kinds—that is, they are not satisfied by sets of objects whose behaviors are homogeneous with respect to causes and effects in a way that finds reflection in detectable natural laws. In fact, our attempts to couch laws in these terms have not only failed, but in some cases have led to the discovery of causally homogeneous classes of objects among the items they designate that can be described by discovered general laws. We do not believe that there is any such thing as phlogiston. Why not? Because hypotheses that appealed to this substance in order to explain phenomena of combustion turned out to be superseded in the course of scientific development by hypotheses that appealed instead to the existence and operation of another kind of substance, oxygen. It was concluded therefore that there was no such thing as phlogiston: the terms did not designate any distinct kind of substance at all. General statements that cited absorption or liberation of phlogiston to explain combustion and its consequences came to be deemed to be false; true singular statements in which particular events were *correctly* cited as the causes of other particular events, but were referred to by descriptions implicitly or explicitly attributing the liberation of phlogiston to objects that took part in them, were accordingly deemed to be misleading, on the ground that although they correctly specified singular causes and effects, they did so by appeal to a concept which designates no real property of things. The hypothesis that the substance oxygen was incorporated during combustion and liberated during reduction enabled us to substantiate the truth of these true singular claims, while explaining their misleading character. They also enabled us to give alternative equivalent descriptions of the particular causes and effects cited in these misleading statements, which showed them to be consequences of and not incompatible with a law that implicitly denied the truth of the general statement about phlogiston that they presupposed. The term "phlogiston" disappeared from the description apparatus of language becuase it did not designate a natural kind, a class of causally homogeneous objects. And we discovered this fact through the failure to frame a well-confirmed hypothesis in which the notion figures.

The ordinary concept of "fish" designates no natural kind either, even though it continues to function in our everyday descriptive vocabulary. As we ordinarily use this term it means something like "a legless aquatic animal," and this includes, along with bass and flounder, sharks, whales, octopus, jellyfish, and starfish. Now it is obvious that this class is biologically heterogeneous, for it includes mammals, whose physiology and behavior are utterly different from members of the phylum Pisces, who in turn are different in behavior and physiology from lower forms of aquatic animals like the shark, or mollusks or crustaceans, or coelenterates. Any attempt to frame a small number of general laws about, for example, the anatomical respiratory mechanism of the class of objects satisfying our ordinary meaning of "fish" is bound to fail. For there is no anatomical mechanism common to whales, trout, crayfish, jellyfish, and coral. Insofar as the class of objects that we normally call fish is not homogeneous with respect to nomologically significant properties, the notion of "fish" simply drops out of biological classification and biological theory. It is replaced by a set of other notions which divide up the class of objects that "fish" lumps together into distinct natural kinds. It is about these natural kinds that simple and precise true universal general statements stand the chance of being discovered. Our guide to this new typology has been our eon-long investigation of the class which satisfies our ordinary notion of "fish" and our discoveries of scientifically significant differences among them. Our continued employment of the term "fish" in the expression of true singular statements (about, for example, which particular object we ate for dinner last night) is a reflection of the fact that for many practical, nonscientific purposes, terms that do not pick out natural kinds are perfectly serviceable, and indeed, dispensable only at the cost of great inconvenience and much irrelevant typological study.

Again, the notion of "race" current in biological and social science does not represent a natural kind; that is, there are no groups of traits common and peculiar to subsets of *Homo sapiens* that figure in general laws that will enable us to make theoretically significant distinctions between them, or to deterministically predict and explain the features and properties of behavior or structure of any individual *Homo sapiens* from a knowledge of his "race." One way of making this point is to say that there are no races. And clearly, what is meant by this claim is not that we cannot distinguish rough classes of geographically located, biologically related groups, but that there are no well-confirmed empirical generalizations about invariable differences between members of these classes. The continued employment of such notions as "race" to make invidious comparisons among individuals and to determine differential treatment of them is condemned as racism and explained as a reflection of irrational and immoral prejudice in part because of our belief that scientific study has shown the concept to be nomologically sterile, to individuate no natural kind.

Similarly, suppose that the notions of 'desire', 'belief', 'preference', and their other mentalistic cognates—'fear', 'hope', 'anticipation', 'expectation', 'want', 'suspicion', 'action'—do not describe natural kinds. That is, suppose that the events, states, and conditions that involve human beings, and that we describe by the use of these terms, do not constitute causally homogeneous classes—sets of events, states, and conditions that can be subsumed under a manageably small number of stateable general laws. If this is the case, then the singular statements that social science and ordinary practice commit us to about the causes and effects of particular mental and bodily states, events, and conditions may be true even though we shall be inevitably frustrated in our attempt to formulate general theories relating these states, events, and conditions. If there are no causal relations in which desires, beliefs, and actions figure, *by virtue of their being* desires, beliefs, and actions, then all of our singular statements referring to them could be true, even though there is no law that refers to them under their descriptions as desires, beliefs, and actions. In documenting the failures of economics, psychology, or the other social sciences to provide specifications for desire, belief, and action independent of the maximization hypothesis that cites these causal variables, we have revealed the failure hitherto to substantiate empirically the belief that desire, belief, and action do constitute natural kinds. The exceptionlessness of the general statement presupposed by these subjects and by our ordinary explanatory practices limits the potential specifications of L's variables to ones which are independent of behavior, and therefore restricts these specifications to neurophysiological ones. But the stupefying complexity, redundancy, and interconnection of brain states finely enough structured to be localized and credited with identity to particular states of belief and desire at particular times, together with the equally myriad bodily movements any one of which could constitute a particular action, make the provision of the specifications required for a mental notion's being a natural kind as unlikely as an empirical possibility can be.

How is it possible for the singular claims of everyday life and social science to be construed as true causal claims about particular reasons and actions, if the terms which we hit upon for describing these causes and effects have no foundations in general laws, do not provide descriptive predicates that figure in nomological relations, do not describe the cited causes and effects under descriptions that reveal their powers to be the causes and effects we credit them with being? The explanation turns on the fact that we can use predicates attributively and referentially, that we can correctly designate the cause or the effect of an event, state, or condition even though we designate it by attributing to it properties which it does not have at all, but which we merely mistakenly believe it has, or believe are its properties causally relevant to its actual antecedents or consequents.[8] Thus, we may correctly pick out a receptacle containing the products of combustion as the one and only one containing (we falsely believe) phlogiston. Of course, we are wrong to attribute to the beaker or test tube the property of containing phlogiston, but we may nevertheless be correct, by accident or not, in our referential claim that one particular item in the laboratory and not any other contains the products of the combustion. Similarly, we may say truly of a beached whale that the large fish died because it was out of the water for too long. Here our causal claim may be true even though both the descriptions employed to make our causal claim and the grounds of the claim are mistaken: mistaken because the animal in question was not a fish, but a mammal, and because, though the lack of water did lead to its death, it was not, as we suppose, through the process of asphyxiation but for some other reasons relating to thermoregulation. Or again, we may causally "explain" a particular person's failure to manifest the genetic sickle-cell trait by noting that he is not a member of the black race. Our use of the notion of "black race" may in fact cut us off from a correct understanding of the causal connection between this genetic trait and persons whose ancestry is traceable back to a spatially localizable area of western sub-Sahara Africa, but our claim is not, for all that, a false one. In these three sorts of cases the singular claims are true, so to speak, by accident. In each case we can correctly cite an individual event's causes or effects because the events we bring together on the strength of false general beliefs, or no general beliefs at all, are in fact related by nomological generalizations that we are ignorant of. And it is because we are ignorant of them, ignorant of the descriptive concepts that figure in them, and that are in fact satisfied by the particular events we pick out, that we wrongly or mis-leadingly characterize the events in the ways we do. Indeed, since most of our ordinary vocabulary for describing the phenomena with which we are ac-quainted has been superseded in the development of science, since very little technical description of even readily accessible phenomena in physics, chemis-try, or biology employs the predicates of ordinary language in their ordinary meanings, it must be inferred that most of our ordinary descriptive-kind terms do not pick out natural kinds: words like "table," "chair," "boy," "cloud," "paper," "metal," "liquid," "light," do not figure in natural laws and do not designate natural kinds. Of course, this does not make the vast number of statements that employ these terms *false*; it only makes their nomological justification difficult, and poses traditional problems for the analysis of causation into nomic subsumption. It is equally clear that since our practical purposes in the employment of these terms are well enough served by such terms in spite of their nomological isolation, we are unlikely to dispense with them except for the explicit explanatory and predictive purposes of science, with its special standards of rigor, precision, simplicity, unity, depth, and systematization.

Now, when we describe a particular event as an action or explain its occurrence by citing a particular agent's mental states in terms that attri-bute to the agents both beliefs and desires, we may in fact be referring to the states which are in fact the causes of the event in question, even though there are no laws relating desires and beliefs to actions, and the laws that do relate the states and the events correctly brought together in the singular statement may be both so large in number and so complex in expression as to make filling out the justificatory details of the explanation practically—indeed, physically—impossible; and these details may be so foreign to our ordinary or social-scientific conceptual scheme as to make the attempt to connect sys-tematically the singular relation in question with other such phenomena theoretically and practically sterile. This in fact will be the case if the neuro-physiological correlates of states of belief and desire are as complex as I have suggested. The quick way to describe such a state of affairs is to recognize that terms like 'belief', 'desire', and 'action' do not pick out natural kinds. For purposes of the scientific study of human behavior, for the descriptive tools needed to uncover laws of human behavior, they are otiose.

The hypothesis that the terms in which we describe and explain human action are not natural-kind terms will explain why we have not found any laws of human action even though we have detected many pairs of particular causes and effects in human behavior correctly. We know from the history of science that entrenched classificatory schemes and misleading descriptive vocabularies have impeded scientific advance as much as or more than the complexities and observational inaccessibility of the subject matter. Equally important, breakthroughs in the discovery of new laws and theories of the most powerful kind have turned on the establishment of new typologies as much as on the provision of new research technologies, more powerful telescopes, energy sources, and so on. The best example of such development is perhaps the Mendeleevian table of the elements, which brought the elements together in new classificatory groupings so suggestive that they provided not only the descriptions needed for the development of the atomic theory of matter, but much of its motivation in the discovery of new elements. Even more simply, reflection on the notion of simultaneity, and on its conceptual position in mechanics, is more responsible for the special theory of relativity than the empirical findings about the speed of light which antedated the theory by some twenty years. Perhaps the obstacle to a social science, to the laws that empiricism demands that we seek but which we have not found, is not to be attributed to complexity and recalcitrance of subject matter but to a well-entrenched typology which reflects no natural kinds and in which, therefore, no laws can be expressed.

Our hypothesis that the explanatory terms of ordinary life and social science are not natural-kind terms renders empiricism consistent with the truth of the singular statements we are constrained to preserve, and makes both of these commitments compatible with the view that though the social sciences have as yet produced no general laws, given a more fruitful typology they are capable of doing so. And it does all this without any implausible appeal either to the comparative complexities of human behavior or to our inability to objectively observe or experiment with it. But despite these evident advantages our hypothesis must meet several significant challenges and objections. The first of these, already noted when our hypothesis was first broached, is the question of the status of L, our general statement connecting actions with reasons. It was denied standing as a law on the grounds that it cannot acquire the theoretical entrenchment whose possibility is a necessary condition for nomological status. We have the potential explanation for this in our hypothesis that reasons and actions are not natural kinds. But how can we square this finding and its explanation with L's truth? It will be recalled that in my argument against L, I allowed that L is a *true* and *synthetic* general statement. Accordingly, I cannot defend L's truth consistently with the claims made about its variables by arguing that it is an analytic proposition, a consequence of the meanings of its variables. Is there any alternative but to admit that L is a law? Yes, for there are many propositions that are true synthetic statements which are universal in form and which are not laws. Such statements are often called accidental generalizations, and it remains a traditional problem for the empiricist to show how such statements are to be distinguished, in ways consistent with empiricist strictures, from nomological generalizations. It is universally admitted that there is a real difference between the laws of physics and chemistry and statements to the effect that every twentieth-century president elected in a year ending in zero died in office or that all the apples in a given basket were Golden Delicious apples. Although statements of the latter kind are true synthetic statements of universal form, they are clearly not laws, and the problem for empiricism is to explain why in ways that do not appeal to empirically inaccessible modalities, for example. Although this problem has not yet been solved, we may still appeal to the accepted distinction and preserve our explanatory hypothesis against any embarrassment produced by L's truth and its synthetic status by classifying it as an *accidental* general statement.

The category of accidental generalization is a convenient pigeon-hole in which to cache L in order to avoid the embarrassment of a true synthetic general statement about reasons and actions, but it is clear that independent grounds must be provided for so characterizing L; otherwise, our argument will be accused of circularity. Moreover, such independent reasons are also needed to explain why terms like 'belief', 'desire', and 'action' do not designate natural kinds. After all, so far our only argument for the claim that reasons and actions are not natural kinds is the failure to discover any laws couched in terms referring to reasons and actions; and our explanation of this latter fact (consistent with the truth of empiricism and the singular judgements we insist on preserving) is based on the hypothesis that these terms do not designate natural kinds. Because the consistency among other independent commitments that is effected by the explanation makes for an asymmetry in the grounds of these two claims, the relationship between them is not plainly circular; rather, it has the character of a highly controversial inference to the best explanation. It is a highly controversial inference because its plausibility rests on commitments to empiricism and because of the existence of an alternative explanation for the same phenomena—the absence of laws—which has not been excluded to its own proponents' satisfaction. Consequently, what is required for a really convincing explanatory argument is some independent reason to accept the explanation, such as one which explicates the accidentality of L independent of our failure to entrench it in a nomological network. If we can give independent reasons for the accidentality of L, we shall also thereby provide independent reasons for denying that beliefs, desires, and actions are natural kinds, and therefore provide a more fundamental, more general explanation for the failure to find laws of human action, one which does not trade on our commitments to preserving empiricism and to our singular judgements.

The explanation that will be provided rests on showing that the terms of L have the character which is distinctive of the predicates of an accidental generalization: they fail to be "purely qualitative," and their meaning involves reference to spatiotemporally localizable particular objects. Having predicates of this character is widely recognized as being a sufficient condition for the nonnomological status of a general statement. For statements that make implicit or explicit reference to particular objects, places, and times do not reflect the sort of universality associated with the notion of "law" and cannot be employed to explain and predict the infinite class of objects to whose existence the meaning of the law commits it. Thus, the death of John Kennedy can hardly be explained by citing the fact that he was elected in a year ending with a zero, even though it is a true generalization that all such twentieth-century presidents died in office. It cannot be so explained in part because the considerations which explain this accidental generalization are of great heterogeneity, and in part because their very diversity reveals the happenstantial, coincidental character of the generalization itself. It is characteristic of general statements which are couched in terms not purely qualitative and designate spatiotemporally particular items that their own explanations are inchoate and unsystematic; and this is, of course, just a reflection of their nomological isolation. Their truth is explained by explaining the truth of each of their finite number of instances, each of which appeals to a different explanatory principle or law. On the other hand, a law *ipso facto* provides an explanation of its instances, and does so both because it follows from more general laws in a direct way not involving appeal to its instances, and because its force transcends the finite number of instances of the law with which we might be acquainted and whose explanation we seek.

One hint that L may have this status of accidental generalization appears in its genesis. For the method by which it was formulated reflects L's contingency on the singular judgements that are its instances. L is formulated by considering particular circumstances and the propriety of everyday objections to proffered explanations of actions by the citation of belief and desire. Its form, and the fact that its antecedent contains just six, and not five or seven, clauses, is a reflection of our merely conjectured inability to imagine other potential objections and the supposition that this inability reflects the nonexistence of further factors. This is one respect in which L lacks nomic force. Moreover, the acceptability of L is in part a consequence of the fact that if further actual or possible particular falsifying circumstances should occur (or occur to our imagination) we may simply formulate a new lawlike general statement just like L which takes our new exception into account by adding the appropriate excluding clause to its antecedent. But this sort of foundation in unsystematic intuition and preservability is the mark of an accidental generalization par excellence. The strength accorded our singular causal claims about reasons and actions is a product of this introspective intuition that through them substantiates L. This immediate self-conscious access to particular states which we have characterized as states of belief and of desire, and which we somehow directly know in our own cases and inductively infer in other cases to be the particular causes of those bodily movements and forbearances that we call actions, has been erected by some social scientists into an entire methodology, under the name of *Verstehen*. The unshakeable recognition of what the particular causal determinants of actions are in individual cases is what confers upon them the inviolability reflected in the common assumption of social science and everyday life, and it demands that an acceptable version of empiricism be rendered consistent with it. It also explains how we can consistently sustain the truth of so many singular causal claims in the absence of any law to underwrite them.

In the next chapter we shall find more than just a hint that L, and any alleged law that mentions desires and beliefs, is an accidental generalization. So far, we have come halfway in our effort to preserve empiricism from untenability, for we have sketched an explanation for the absence of laws in the social science that does not merely reflect lack of industry or genius, that makes sense of four generations of empirical search for a law of human action, by showing that the search was not conceptually misconceived, but was bound to fail because of a commitment independent of empiricism. That commitment determines a typology which frustrates any attempt to discover general laws and which therefore must be surrendered in the description of human events that can find its way into such laws. But just because the description must be surrendered for nomological purposes, it does not follow that we must surrender our confidence in the occurrence of the events which that typology has been used to describe. And this is an important concession for the defense of empiricism as a prescription in social science. For without it, the adoption of empiricism could be claimed to have consequences so intuitively implausible as to make it practically inconceivable. But, as I have said, we have only come halfway in our defense. Although we have found a way to pass formally between the dilemma of an implausible defense of empiricism and the embrace of an even more implausible nonempiricist philosophy of social science, we must execute the actual passage, both by substantiating independently the claim that reasons and actions are not natural kinds, and by showing exactly what sorts of laws empiricism can lead us to expect in the social sciences. For it will be a hollow defense of empiricism as the correct method in social sciences if the result merely renders this philosophy of science formally consistent with the existence of a nomological wasteland, instead of showing how the concerted application of such a philosophical commitment can provide generalizations for subjects hitherto lacking them.

6

Human Kind and Biological Kinds

Many philosophers have maintained that mental predicates are connected by their meanings with the particular appearance and, indeed, physiognomy of human beings, of members of the species *Homo sapiens*. The peculiar non-scientific behaviorism of Ludwig Wittgenstein's treatment of the use of mental terms is the most influential example of this attachment philosophers find between thinking and being a human being. In his exposition of Wittgenstein's aphoristic expression of this view, Norman Malcolm has written, "Since it has nothing like a human face and body, it makes no sense to say of a tree, or an electronic computer, that it is looking or pointing or fetching something (of course one could always *invent* a sense for such expressions). . . . Things which do not have the human form, or anything like it, not merely do not but *cannot* satisfy the criteria for thinking. I am trying to bring out part of what Wittgenstein meant when he said, 'we only say of a human being, and what is like one that it thinks,' and 'The human body is the best picture of the human soul.' "[1] It is in general hard to understand the views of those who would deny mental states to computers whose behavioral repertoires are identical, under stringent conditions of duplication, to those of human beings, except by attributing to such writers the view that at least some conscious states are either definitionally, or at least empirically, inexplicably limited to members of the species *Homo sapiens*, and perhaps to some specially trained members of closely related primate species.[2] Even the proponents of machine consciousness seem to accept this doctrine, to the extent that they recommend redefinition of mental predicates in the face of their successes in simulation. Insofar as no simulation or duplication of human behavior by a machine can logically oblige us to accord the machine mental states of the kind we accord ourselves, the meanings of the terms employed to pick out these states must be given in some species-specific manner. To the extent that the question of whether machines can be accorded these states remains intelligibly open, then no matter how idle for practical purposes the question may ultimately become, the definitions of these states must make appeal to the assumption that the states are at least paradigmatically or exclusively states of men, women, children and a few well-trained chimpanzees. Even if we eventually accord mental states to computers, or to apparently sentient nonterrestrial creatures, it will be on the basis of an analogical argument from our own cases. They will be accorded states of desire and beliefs, and their actions will ordinarily be explained by the citation of a general statement like L only because of the similarity of their behavior to that of the "benchmark" behavior of *Homo sapiens*. Of course we can and do accord nonhuman systems purposive or teleological states independent of any analogical appeal to our own behavior and its determinants. But consciousness involves more than mere teleology, or else we would have to accord it unexceptionally to many complex but purely physical systems like thermostats and inertial guidance systems. It is this extra burden of human behavior, beyond its goal-directedness, that we appeal to conscious states in order to explain, and which philosophers frequently describe in terms of intentionality or content of consciousness. Our difficulty in saying what would constitute a machine's having the desire that or the belief that a particular proposition be true accounts for the resistance to according such states to machines. For we know that humans have such states, although we seem equally perplexed about what exactly constitutes such states in our own cases. Perhaps because we cannot account for the undeniable intentionality of our own states, we can only accord such states to nonliving systems on the basis of a strong analogy between their behaviors and ours, that is, by implicitly defining such states in terms of a paradigm of behavior characteristic of *Homo sapiens*. It is no surprise that those, like Wittgenstein, who deny that we attribute mental states to other human beings on analogical grounds because such grounds are too weak to sustain our moral certainty about these attributions, are the very same philosophers who refuse to accord mental states to nonhuman items on the same ground, that the analogical basis is insufficient to sustain the conclusion that nonhumans have intentional states. The argument between these philosophers and those who would accord mental states to machines and to other nonhumans is in large measure one about the prospects for an analysis of intentionality that will free it from conceptual connection to the concept of a human being, or about the conditions of adequacy that an analogical argument of the required kind will have to satisfy. Both parties to the dispute agree that at present we have no way of separating the meanings of mental terms from the notion of *Homo sapiens*.

For our purposes this fact is important, because, as I shall argue, '*Homo sapiens*' does not name a *natural kind*, and is not a predicate at all, let alone a qualitative predicate. Rather it is a *name*, a proper name for a discrete, spatiotemporally bounded particular *thing*. As such, it is no more likely to figure in a general law than 'Napoleon Bonaparte' or 'Kalamazoo, Michigan' or 'Mona Lisa a Gioconda.' Although the three items mentioned behave in accordance with general laws, the terms employed to mention them do not figure in the expression of laws, nor do any other predicates defined by appeal to these terms. This is a consequence of the requirement that laws contain only qualitative predicates, that their terms neither explicitly nor implicitly mention particular places or things.[3] But if '*Homo sapiens*', and all other species terms, like '*Canis familiaris*', '*Cygnus olor*', are names of spatiotemporally restricted particular things, then neither they nor any term defined explicitly or implicitly by appeal to them will be permissible in general laws either. And of course if notions like 'desire', 'belief', and 'action' are to be defined, as suggested above, either essentially or paradigmatically as *human* attributes, then they too will be excluded from general laws as failing to be of the purely qualitative sort which nomological generalization requires. And any general statement which employs them will, if true, be at most an accidental generalization. But this conclusion will give us the independent explanation of why *L*, though true and exceptionless, is no law after all, and will provide just the noncircular explanation of the failure to find laws of *human* behavior which we require: one consistent both with the dictates of empiricism and with the truth of that vast body of singular causal statements that we express, albeit in nonqualitative terms. Moreover, further reflection on what sort of a term '*Homo sapiens*' is will reveal the character of the lowest-level laws which we can expect to discover about the behavior of human beings. Our first task in pursuing this line of argument is to show that, in general, species terms do not name natural kinds, but particular things, and this will involve an essential excursus into biological theory.

Our task is made more difficult by the fact that throughout the history of logic the names of species have been invariably treated as the names of classes or sets of indefinitely large numbers of individuals. They have been treated as predicates which figure in open sentences like '*x* is a swan' that can be true of any number of individual objects from zero to infinity. Species names were initially assigned to distinct kinds of individuals, and much science (natural history) consisted exclusively in consigning individual organisms to their appropriate species. Species themselves were and commonly continue to be assumed to be fixed and utterly distinct in their membership, even though in most cases taxonomists have never been able to provide sets of necessary and sufficient conditions for an organism's being an example of a given species. Moreover, there are remarkably few generalizations of exceptionless sorts or with any counterfactual force about the species isolated in taxonomic research. 'All swans are white' may serve the philosopher as a generalization for purposes of introductory discussion, but it plainly has exceptions that deprive it of nomological status. In fact, some arguments, to the effect that biology possesses no distinctive laws of its own, trade on the inevitable exceptions and restrictions with which the biologist must hedge around his claims about general features of members of any given species.[4] As we shall see, such arguments reflect mistaken notions about the subject matter of biological laws: the general findings of biology and biological theory do not bear on regularities about particular species, like *Canis familiaris* or *Didus ineptus* (the dodo); they bear on laws with which all species are in conformity. The subject matter of biological theory is the behavior of any species, not of particular ones, just as the subject matter of mechanics is the behavior of any body, but not any particular one.

Of course, the most important theory of biological science is Darwin's theory of natural selection. This theory tells us that as a result of hereditary variation among the members of a species and selection over this variation, *species evolve*. Subsequent work revealed that there was a special unit of heredity whose behavior accounts both for variation through mutation and inheritance through replication: the gene. The individual organism is, of course, the immediate unit of selection; it is the item on which environmental pressures operate, and whose survival and reproduction determine the character of subsequent members of the species. The crucial feature of Darwinian theory is that the unit of evolution is the *species*; it is they which evolve, and their evolution consists in changes in the relative proportions with which their members in successive generations manifest the varying hereditary characteristics, or phenotypes, determined by changes in the units of mutation and in the units of selection. This explains both why biologists can provide no necessary and sufficient conditions for various particular species, and why there seem to be no exceptionless generalizations about particular species. Since species evolve, there is no trait which jointly meets the requirement of being hereditary and therefore essential to the species in the way required for defining the particular species, and at the same time either necessary or sufficient for being a member of the species. Similarly, the apparently general statement that 'all polar bears are white' is not a lawlike consequence of evolutionary theory, although if true, its truth can be explained by citing generalizations about the existence of adaptive variations in any species which is exposed to generally described environmental conditions and which survives in those conditions over the long run. Since, according to the theory of natural selection, species evolve, it follows that they should be treated, not as classes whose members satisfy some fixed set of conditions—not even a vague cluster of them—but as lineages, lines of descent, strings of imperfect copies of predecessors, among which there may not even be the manifestation of a set of central and distinctive, let alone necessary and sufficient, common properties. A kind or sort for which *no* properties whatever could be definitional is no kind at all, and a kind which remains unchanged while any or *all* of the defining properties of its instances change over time is equally hard to comprehend. But neither of these difficulties arises for a particular object which may change its properties over time, may evolve, and about which the question of defining properties does not come up. More importantly, biological theory dictates the accordance to species of an individuation and a unity which is unintelligible except on the assumption that species are particulars. Thus, although the disappearance of all atoms of atomic number 79 does not entail the disappearance of a space in the periodic table of the elements, but only its temporary emptiness, the disappearance of all members of a species entails its extinction, the utter disappearance of the species; the appearance of new organisms qualitatively indistinguishable from the extinguished species' members does *not*, in biological theory, constitute the *reappearance* of the *same* species, but represents an entirely new one, just because it did not arise in any line of descent from the old one (which, having no issue, became extinct). Moreover, just as individual organisms, which are themselves particulars, may undergo vast changes in genotypic and phenotypic properties, may divide into two or fuse with another into one, may continue to exist while generating new individuals, so too, evolutionary theory requires that species be capable of all these things. Thus, as binary fission makes two particular organisms out of one, geographic or other environmental isolation can make members of the same lineage reproductively isolated after a long enough period of separation, so that whether their appearance is similar or not, they must be classed as *new* and different species. Indeed, this is how, on the view of some biologists, speciation proceeds. Where genetic change makes both branches reproductively isolated from the original lineage, the original species may be said to become extinct. When substantial continuity and the possibility of reproduction between members of the original lineage and one of the subsequent lines remains, the phenomenon is akin to that of an individual generating a new individual while continuing to exist as a unit. Similarly, introgression may obtain between two species, sometimes creating a third or interconnecting the lines of descent constituting each species. Now, although examples of one or more kinds of things may do any or all of these things, the kinds themselves cannot. As Plato so clearly recognized, kinds are immutable; it is only things which change. Individuals are spatiotemporally bound; they are discrete entities with a location and a history, even though it is sometimes difficult to plot their boundaries and their beginnings or endings. So too with species. Temporal boundaries among species are marked by extinction, or by the production of sterile offspring, or by the onset of reproduction among an isolated small group of founder organisms, or again by the process of polyploidy, in which miosis—the multiplication of genetic material—occurs without cell division, doubling and thereby changing the chromosomal material that determines reproductive possibilities. Spatial boundaries are simply given by the distribution of members of the species. Spatial continuity is reflected in reproduction and in other sorts of species-specific behavior. Naturally, species do not have all the unity and spatiotemporal continuity of the usual example of a particular object, like a table or a chair, but they certainly have enough to be so classified, when we consider the vagaries of individuation for such particulars as nations or cultures or even organisms with peculiar biological potentials. Most important, species have the uniqueness that is necessary for being individual items and sufficient for not being general classes or kinds of items. Qualitative similarity up to any degree of completeness is neither necessary nor sufficient to determine whether two organisms are members of the same species (as it should be were species kinds or general classes); what is necessary for such determination is that the two organisms be links in the same spatiotemporally restricted chain of genetic inheritance. Without such a connection between the two organisms, there cannot be an evolutionary path between them, and since species are the units in which such paths are laid out, the organisms cannot be members of the same species.[5]

The conceptual status of species names as names of spatiotemporally extended particulars is reflected and reinforced in the character of biological laws, empirical generalizations, and statements of accidental universality or finite scope. Thus, the apparently general statements that all beavers build dams, even if true about each and every beaver, or that all swans are white, do not count as biology's candidates for nomological respectability, nor are they even empirical generalizations of that subject. For, properly understood, they are singular statements which predicate properties to the components of a spatiotemporally restricted particular. The supposition that species are kinds gives such statements the appearance of universality in form, while the existence or physical possibility of exceptions as well as the restriction of their domains seems to deprive them of nomological force. That is why they are sometimes treated as accidental generalizations, and why more than one philosopher of science has denied that biology has any distinctive laws, and has asserted that it simply applies nomological generalizations from the "harder" sciences. But not only are such statements not the *laws* of biology, they are not even its *rough empirical generalizations* (general statements with exceptions, but with enough nomological force to permit their explanation by the real laws of biological theory). Among the empirical generalizations of biology are the following statements, which mention no species whatever, but are general claims about all species which are roughly correct:

Unspecialized species tend to avoid extinction longer than specialized ones.
Body size tends to increase during the evolution of a species.

Contemporary species living in the same environment tend to change in analogous ways.

In colder regions the members of warmblooded species are larger than members in warmer regions, though their extremities are proportionately smaller.

These are examples of the *lowest*-level generalizations of biology; they manifest exceptions, like other empirical generalizations, but these exceptions can be explained by appeal to the *same* more general claims that help explain the exception-ridden generalizations themselves. And these higher-level claims are the laws of biology. These empirical generalizations, unlike statements about swans and beavers, can be expected to obtain, *ceteris paribus*, on other life-supporting planets, and indeed on planets which manifest "life" that is so different from our own as to be unrecognizable to the anthropocentric eye. If statements like these are the lowest-level generalizations of biology, then the natural kinds of biology cannot be the particular species that the taxonomist constructs, for these laws mention no particular species. Rather, they mention the kind that every particular species is an example of, by virtue of having the property of being a species. If particular species like *Ursus ursus* or *Canis lupus* are not natural kinds, since they do not figure in even the lowest-level generalizations of biology, what alternative is there for them besides being treated as names of particulars? We cannot treat these terms as we might treat "phlogiston," as terms which denote nothing and whose use in singular statements is at best misleading and at worst nonsensical. Species names are not kind-terms which have been superseded in the course of scientific advance. They do enable us to usefully individuate and relate the vast number of individual organisms that populate our planet, and they enable us conveniently to show the hereditary relations between various organisms, their evolutionary distance from one another, and the kinds of environmental factors that make for differences between them. How can we retain the advantages which the use of species terms provides, consistent with the recognition that they do not reflect natural kinds? Certainly not by according them the status of nonnatural kind, like the nonnatural kind "table." Rather, we may accomplish this by treating species names as names of things instead of kinds, natural things which are named in an organized way that reveals much systematic, biologically fruitful information about them while recognizing their particularity and individuality.

If a statement about the color of swans is not even an empirical generalization, and the claim that body size increases through the evolution of a species is no more than such a generalization, what are examples of biological *laws*? Good examples are provided by the Mendelian laws of genetics and the laws of segregation and independent assortment of genes. But these laws, like many others in functional biology, are well on their way to reduction to biochemical and chemical laws, and are consequently beginning to lose their claim to distinctively biological status. More autonomous, and for our purposes, more important examples of biological laws are provided by the principles of evolutionary theory. Such principles are especially significant in the present context, both because their character further establishes the status of species concepts and because we know that these laws govern human phenomena as much as they govern the behavior of any other species, so that the laws of human behavior must minimally be consistent with them. Consider the following four general statements, which, it has been argued,[6] axiomatically represent the content of the theory of natural selection:

1. There is an upper limit to the number of organisms in any generation of a species.
2. Each organism has a certain quantity of fitness with respect to its particular environment.
3. If D is a physically or behaviorally homogeneous subclass of a species, and is superior in fitness to the rest of the members of the species for sufficiently many generations, then the proportion of D in the species will increase.
4. In every generation of a species not on the verge of extinction there is a subclass, D, which is superior to the rest of the members of the species for long enough to ensure that D will increase relative to the species, and will retain sufficient superiority to continue to increase, unless it comes to constitute all the living members of the whole species at some time.

Unlike the merely empirical generalizations mentioned above, these principles are supposed to be exceptionless, although there is difficulty giving specific content to the notions of "sufficiently many generations" and "long enough time to ensure increase." The important points are again that these biological laws mention no particular species, and that they treat species as particular evolving lineages, not as types or classes or kinds of organisms. We cannot deduce anything about the evolution of any particular species from these laws, nor can we predict anything about the future of a given species from these laws *alone*. Some philosophers have criticized evolutionary theory on this score, and even condemned it as an empty and unfalsifiable account bereft of explanatory power; other philosophers have moved in the opposite direction on the tracks of the same argument and claimed that since evolutionary theory makes no predictions, and since it is clearly a theory of great explanatory power, it must be false that there is any parallel between explanation and prediction, contrary to empiricists' views. Both sorts of philosophers' complaints reflect a misunderstanding of evolutionary theory, and of the status of claims about particular species. They mistakenly suppose that statements about the properties of species or their members are general statements, and that therefore they should follow from a theory about the evolution of species in general. Since the statements that the theory alone enables us to infer about particular species are either true because (close to) tautologous or so vague as to be unassessable, some philosophers conclude that logical empiricist strictures on scientific explanation are wrong, for this clearly explanatory theory makes no specific predictions. Other philosophers argue that because such methodological strictures are correct, evolutionary theory is a scientifically disreputable enterprise. In fact, the relationship between the theory and particular statements about special species sustains neither of these two views. The theory cannot be expected to issue in the sort of singular statements that are open to test, without the addition of statements of initial conditions, and yet species-specific statements are of just the former sort. On the other hand, with the provision of the required statement of initial conditions, the theory and its derivative laws will enable us to make predictions up to the levels of accuracy of the statements of initial conditions supplied. For example, from more complex mathematically expressed versions of the four laws of evolutionary theory enumerated above, the following general statement is derivable as a theorem:[7]

If D_1 and D_2 are distinct species within an environment of cohabiting organisms, D, and if D_1 is superior to D_2 as long as it constitutes less than e, a certain portion of D, while D_2 becomes superior to D_1 when D_1 comes to constitute more than that proportion of D, then the proportion of D_1 to D will either stabilize at e or oscillate around e.

Now if D_1 and D_2 are two species related as predator and prey, then our derived law will explain why their populations oscillate around fixed values. And given two particular species, say caribou and wolves, if it can be independently shown that the wolf population depends for existence solely on predation of the caribou, and that survival of caribou depends solely on avoiding the predation of wolves, the theorem tells us that the average fitness of wolves will decrease as their ratio to caribou increases and thus their survival rate decreases, so that there must be a balance between the numbers of these species. In other words, if we can establish these initial conditions in the interrelations between wolves and caribou, the theory will enable us to predict that each of them will eventually exist at or around a fixed level of population, regardless of what population distribution the species began with.

Of course, the biologist, like other natural scientists, is not content with such a generic prediction, with the prediction of the existence of an equilibrium value of population for competitor species; he would like a prediction of the actual value where that value is unknown, and an explanation of that value where it is known. The method of acquiring such predictions and explanations in biological theory has consisted in the construction of a series of mathematical models of increasing realism and sophistication, with a better and better fit both to actual data and to the factors that theory tells us determine fitness and populations. The earliest of these models was developed by A. J. Lotka and V. Volterra,[8] and took the following form:

$$\frac{dH}{dt} = (a_1 - b_1 P)H$$

$$\frac{dP}{dt} = (-a_2 + b_2 H)P,$$

where H is the size of the predated or host species, and P is the size of the predator or parasite species, a_1 is the net growth rate per individual of the host or predated species in the absence of predation, and b_1 is a parameter measuring the predation rate per individual of the predator species. Our derived law tells us that there are values for H and P above which their derivatives with respect to time are zero, and this restriction together with the values of the parameters will enable us to generate a precise prediction from evolutionary theory about the population level of given species. This early model, of course, fails to take many evolutionary forces into account, and subsequent models attempt to correct this lack of realism in several directions. For example, we may add variables to the model in order to account for the presence and degree of available cover for prey, or where predation is a function of the food needs of predators instead simply of the density of prey. Alternatively, more general models have been devised, and these have been compared to such natural predator-prey systems as house sparrows and sparrow hawks, muskrat and mink, snowshoe hare and lynx, mule deer and cougar, white-tailed deer and wolf, moose and wolf, and bighorn sheep and wolf.[9] Since the number of such models is potentially very large and the data do not always enable us to make a decisive choice between them, it is a matter of interest to know what the general constraints are on these models by virtue of the law, derived from evolutionary theory, that phenomena modeled must reflect stable equilibria. This too has been a subject of theoretical interest, and it has been argued that the four following assumptions are necessary and sufficient for the required equilibrium:[10]

1. Increasing size in parasite or predator species reduces its own and its host's or prey's growth rate.
2. Increasing size in prey or host species reduces its own growth rate, but increases that of the parasite or predator.
3. For both species there are minimum sizes at which they both have positive growth rates in all circumstances.
4. Each species has a maximum size at which its growth rate sinks to zero.

One feature of this aspect of biological theory that is of special interest in comparison with work in social science is that its concern for the general conditions of population equilibrium, and for the production of proofs of the existence of equilibrium, mirrors the traditional concern of economics for the specification of conditions of general equilibrium in economies, and for the proofs of the existence of this equilibrium. That both economics and biological theory should search for such conditions, and manifest interests in such proofs, of course reflects the extremal character of their most important theories, utility maximization and natural selection, respectively. The striking difference is that biological theory embodies a much more well-confirmed set of theoretical assumptions—the four axioms of evolutionary theory—and partly in consequence, provides not only formal conditions for general equilibrium but the actual numerical values of population levels to reasonable degrees of accuracy for a variety of actual species at which this equilibrium is reached. All this shows that biology's general assumptions are far better candidates for nomological standing than those of economics or any of the social sciences, and that the concepts that figure in its assumptions stand a far better chance of being natural kinds than do those of contemporary social science.

Like the economist's models, the biologist's models make highly unrealistic assumptions. Thus, consider the following list[11] of assumptions actually employed in and characteristic of models constructed to account for the relationships among different species:

a. The species under investigation occupies a spatially homogeneous environment and conditions are temporally constant.
b. The system is closed, so that interacting species are not reinforced by immigrations or depleted by emigration.
c. Each species responds to changes in its own and others' sizes instantly, without delay.
d. Variations in the age structure of the members of species do not occur or can be disregarded.
e. The interaction coefficient between each pair of species is unaffected by changes in the species composition of the remainder of the community.
f. The genetic properties and hence the competitive abilities of a species are independent of its size.

Not only will the economist be comfortable with such unrealistic assumptions, but he will recognize some as his own and the rest as parallel to his own. Thus, *a* and *b* are assumed in conventional economic theory; *c* is assumed with respect to decision units and markets; *d* is assumed to hold for all consumers; and *e* and *f* have their duals in assumptions about returns to scale and substitution effects among productive factors. This parallel is highly significant, for it shows that the failures of economic theory to provide predictive results at least as well confirmed and practically useful as those of theoretical biology cannot be attributed solely to its employment of unrealistic assumptions. For these same assumptions figure in the construction of models that we know to be predictively fruitful, and have all the explanatory power of the well-established general theory that they are derived from. The failures of economic theory must be sought at least additionally, and perhaps exclusively, in the character of its fundamental general statements, instead of in its unrealistic boundary conditions. We know that every one of the six assumptions enumerated above is false and that each of the causal forces whose efficacy they deny actually does help determine the value of the level of population towards which the size of a species tends, so that the models constructed with these assumptions can only provide estimates of this value. But we also know, independently of the models, and of the assumptions, that there is such a value. That is, we can be as certain about the existence of such a value as we are of the general laws of the theory of natural selection from which its existence is deducible. By contrast, we cannot say whether the equilibria levels of prices and production generated by the economist's models, with assumptions like a through f, are estimates of a value that we know independently to exist because we do not have the same degree of assurance about the assumptions of maximization that play for the economist the role which natural selection of the fittest plays for the biologist. This is a parallel to which we shall return, but for present purposes it is important to note that in biological theory particular species play the same role that individual agents, markets, and industries play in economic theory. Thus, just as economic theory names no agents, firms, markets, or industries, biological theory names no species. Just as economic theory purports to explain the behavior of instances of the kinds agent, market, and industry, so biological theory purports to explain the behavior of the kind species. Just as we would not expect a generalization about General Motors or IBM to follow from the central principles of economic theory *alone*, neither should we expect to find a generalization about kangaroos or lemmings to follow from evolutionary theory alone. And just as we would not expect an exceptionless statement about Imperial Chemical Industries or all of its employees to figure as a law of nature, derived or not, we cannot expect a statement about giraffes' necks or elephants' trunks to be laws either, no matter how exceptionless their claims are about members of these species. To see all this, one may again turn to the assumptions enumerated above and to their parallels in economic theory. If species were kinds instead of spatiotemporally extended and limited particulars, the parallel between these assumptions and the assumptions economists make about spatiotemporally extended particulars like agents and firms would not be so manifest.

We may conclude from this long excursion that we cannot embrace the truth of the theory of natural selection, in its present character at least, and continue to treat particular species as kinds. Of course, our venture into biological theory has touched on other matters as well, and has involved discussion of matters that will have further ramifications for the argument of this book; otherwise, so extended and technical a discussion would have been out of place. Nevertheless, it is crucial that we be strongly convinced that "*Paramecium causatum*," "*Drosophila melanogaster*," and "*Homo sapiens*" are not purely qualitative predicates but name spatiotemporally restricted particulars. It immediately follows from this that we can expect no laws of paramecium behavior, or fruitfly behavior. This is no surprise, for no one has ever expected that there are regularities in nature that are manifested exclusively and uniformly by each and every paramecium or fruitfly. No one has ever expected to detect generalizations about all and only beavers or sunfish, orangutans or baboons. Had such expectations arisen, semi-autonomous sciences of beaver behavior, say, might have appeared, spawned their own journals, launched arguments for their substantive and methodological autonomy, and pursued research programs calculated to establish a systematic hypothetico-deductive theory of beaver behavior and an estimate of the values of the parameters that the theory claims to limit the behavior of members of this species. Not only is the very description of such a science preposterous, but the serious labors of ethologists who study small numbers of one and only one species clearly reflect the fact that among biologists there is no misunderstanding of the status of ethological findings as constituting an autonomous body of nomological generalizations about members of a natural kind, or even preliminary steps to such a body of general laws. And yet the prospects for a science of human behavior, for a body of nomological generalizations manifested exclusively and uniformly by the behavior of each and every member of the species *Homo sapiens*, are no greater than they are for a science of beaver behavior. And if human behavior is the subject matter of the social sciences, then their claims to substantive and methodological autonomy from the natural sciences and from one another, their research programs devoted to the production of laws of human behavior, or to its economic, or sociological, or indeed psychological aspects, are just as vain as those of a science of beaver behavior.

We may express this claim in terms of the problems of Chapter 5, their solutions, and the limitations on these solutions there noted. There it was claimed that we may formally pass through the horns of the trilemma posed by empiricism's demand for laws of human action, their absence in the social sciences, and the truth of most of our singular judgements about the causes of particular actions, by the hypothesis that the kinds in which we classify the causes in the singular statements are not natural ones. Therefore, it is to be expected that no law relating them to actions will be forthcoming, even though the singular statements are true (they correctly pair actions with their causes, misleadingly described), and even though there are laws (expressed in terms of kinds utterly different from those of common sense) underlying the singular statements. The circumvention of the trilemma proposed was described as formal, because in and by itself it offers what may at best be described as a mere logical possibility. Without some independent evidence for the truth of the hypotheses that underwrite this purely formal, logical possibility, we will lack the conviction that it represents more than such an abstract possibility.

Our excursion into biological theory has provided some of the requisite independent evidence, for it has shown that distinctive laws of human behavior are impossible because they would be about the members of a spatio-temporally restricted particular and would have to be expressed in terms that do not pass the test of being purely qualitative. Yet we have also seen that there are some laws governing human behavior not couched in ordinary terms: minimally, the laws of the theory of natural selection which describe some aspects of the behavior of any species and its members, including *Homo sapiens*. Thus, we have independent evidence that human behavior is law-governed, but that there are no laws of distinctively human behavior, as opposed, say, to mammalian behavior or simply animal behavior. Our excursion provides more evidence for our solution to the trilemma than this, and in fact converts it from the formal solution of a philosophical puzzle into the outline of an independently grounded explanation for the failures of the social sciences.

Recall that in Chapter 5, *L*, our exceptionless general statement relating actions and their reasons, was denied nomological status because the isolation of its constituent terms from correlation with indubitably natural kinds precluded *L*'s nomological entrenchment and consigned it to the status of accidental generalization. Now we can see, again independently of our formal solution to the trilemma, that *L* must indeed be at best an accidental generalization. For *L* asserts the co-occurrence of states of members of the species *Homo sapiens*, states which can only be defined in terms of their exemplification by members of this species, that is, by components of a spatiotemporally restricted particular individual. In fact, this result, that *L* must be an accidental generalization, is independent of the argument that its exceptionlessness coupled with the complexities of neurophysiology jointly insulate it from theoretical entrenchment. For this conclusion now can be seen to turn on recognizing the character of species names and embracing the view that our only current characterizations of the states, events, and conditions mentioned in *L* make inevitable reference to a particular species. If we accept these claims, not only does it turn out to be no surprise that the kinds of events, states, and conditions *L* appeals to are not correlated with a small number of natural kinds described in the biochemical language of neuro-physiology, but it would be a surprise—indeed, something little short of a miracle—to discover that there are a finite number of states that jointly satisfy a set of characterizations which are spatiotemporally restricted and another set which do reflect natural kinds. A correlation between classes of states of belief and desire on the one hand and a causally homogeneous class of neurophysiological states would be as surprising as the discovery that each and every winner of the Irish Sweepstakes lottery had an additional chromosome beyond the forty-six we are normally endowed with. Either case would be a puzzle worthy of considerable research. In effect, our discovery about the conceptual status of the term *Homo sapiens* provides us with an independent reason to believe that L is an accidental generalization at best, and with an explanation of why it cannot be nomologically entrenched that is equally independent of the fact that so treating it enables us to circumvent the philosophical and methodological trilemma facing us.

Of course, there are laws governing human beings and their behavior—for example, the laws of the theory of natural selection. And there are laws governing the relationships among the particular events, states, and conditions which the antecedents and consequent of L refer to: the laws of physiology and neurophysiology, of biochemistry, chemistry, and physics. But L is not deducible from these laws, nor will L enable us to deduce any further nomological generalizations from these laws, because L implicitly mentions a particular thing with restricted spatiotemporal location: the species *Homo sapiens*.

The explanation of the failure of the social sciences to have as yet uncovered any general laws need not, therefore, involve the inapplicability of empirical methods in the study of human behavior, nor the denial that most of our singular statements about it are true, nor the greater complexities and observational inaccessibility of the research object of the social sciences. The fault lies with a mistaken belief that the truth of the singular statements, which turns on their successful *reference* to the causes of human action, implies the truth of some nomological generalization which grounds the causal connections on the manifestation of the properties *attributed* to their relata by the singular statements. This is the hypothetical that motivates the search for a law of human action. We may surrender it without surrendering its antecedent, and without surrendering the search for laws of human behavior. We need only recognize that such laws will not express connections between beliefs, desires, and actions, as we ordinarily describe them. How will they express the connections reflected in our true singular statements? This is in effect the question: What are the natural kinds under which the causes and effects described in our ordinary terms really do fall? As such, it is a question that can be answered neither by philosophical analysis nor by empirically untested speculation. The natural kinds under which the particular pairs of causes and effects fall can only be read off the laws of nature in which they figure. Since our own assurance about any of the lawlike statements we embrace is at best inductive, we shall never acquire any greater certainty about just what are these natural kinds into which particular reasons and actions fall.

Of course the physicalist thinks he knows the most *general* natural kinds into which states of human beings and their movements fall: these are just the natural kinds which can be read off the laws of physics, and they subsume the kinds of events, states, and conditions in which humans figure, just because humans are nothing but complex physical systems. Moreover, by virtue of his commitment to the actuality of the reduction of chemistry to physics, and to the potentiality of reducing biology to chemistry and eventually to physics, the physicalist thinks he knows not only the most general natural kind into which human behavior and its determinants fall, but also the narrower kinds reflected in the distinctive laws of chemistry and biology. For human beings are not only physical systems, they are more specifically complex structures of organic chemical material that satisfy laws governing all biological species. The interesting question for the physicalist is not *whether* there are natural kinds into which human behavior falls, or what some of these kinds are, but rather what is the *narrowest* set of kinds which subsume this behavior; in other words, what are the lowest-level laws we can expect to find in our search for the explanation of human behavior. This question is not merely interesting. Answering it is crucial to our expectations for and standards of systematic explanation, reliable prediction, and effective control of human behavior. If the lowest-level laws governing the behavior of people were the laws of physics, if the narrowest natural kinds under which states of belief and desire and the events we call actions are subsumed are just those that can be described in terms of predicates from mechanics, electromagnetism, etc., then we cannot expect to be able to explain or predict the occurrence of a particular human action any more fully than we can explain or predict the movements of a ball on a roulette wheel. We can explain why the ball landed on 28 *ex post facto*, by attributing values of mechanical variables to the ball and the wheel at some prior time, and then sketching out how these initial conditions, together with the laws of physics, determined the final location of the ball on the roulette wheel. And although in principle there is no limit beyond those of quantum mechanical uncertainty to the exactness of our explanation, in fact we could not have predicted the result by any means that would be of *practicable* use by a gambler. That is why the roulette wheel is used for gambling. Similarly, if the lowest-level laws governing human behavior were the same ones that govern the behavior of roulette wheels and balls, our explanations of particular human actions could not be expected to be any more than schematic, appealing to the operation of the laws of physics on unspecified values of position and momentum of bodies, with unspecified coefficients of elasticity or tensile strength or thermometric expansion, and so on. And our predictions would require descriptions of initial conditions so esoteric in their demands on physical knowledge and so variable from individual to individual that they would be without practicable significance for any purpose currently envisioned by social scientists and those who employ their findings and theories. And the case would be little different if the narrowest natural kinds subsuming the states, events, and conditions in which humans figure were those of some highly restricted branch of organic chemistry.

It is the desire to explain in more than merely schematic, hand-waving fashion and, even more important, to be able to predict and control to within practically useful limits the behavior of individuals and groups that leads to the search for laws governing this behavior that are of a far lower level than those of a restricted branch of organic chemistry. The failure to find such laws at the level of belief, desire, and action which the history of social science reflects, and which the conceptual dependence of these variables on a particular species explains, rules out generalizations embodying kinds as narrow as those of belief, desire, and action. A certain amount of work by social and behavioral scientists can be understood as reflecting a recognition of this state of affairs. For example, the research program of operant conditioning associated with the name of B. F. Skinner is clearly one which eschews beliefs, desires, and actions as the kinds about which laws of human and nonhuman behavior can be formulated. Similarly, the attempt to formulate an account of human behavior by inference from successful computer simulations of salient aspects of this behavior can be construed as the attempt to find practically useful terms, less species-specific than "reason" or "action," in which to describe the explananda phenomena, terms which we have some independent reason to believe figure in laws of nature that can be independently established. In the case of Skinnerian conditioning, the laws are alleged to find confirmation in the behavior of all living systems beyond the simplest. In the case of computer simulation, the warrant for the generalizations is to be found more speculatively in information science, and ultimately in our physical understanding of computer simulation of human functions. Consideration of these two research programs in the light of the findings of this chapter will reveal the issues on which their success must hinge, and will also reflect the significance of the claims of this chapter.

Explanations of human action that employ principles from operant learning theory are, like those that figure in ordinary contexts, teleological in form; they explain the occurrence of an event by appeal to its effects on the organism. Unlike the teleological explanations of common sense, however, the forces to which they appeal are not characterized in ways that require them to be conscious states of human beings. Accordingly, the explanatory principles of operant conditioning theory are as applicable to members of any other species as they are to our own. In fact, the principles were first formulated with respect to the behavior of laboratory animals, and only extended in their application to human action by virtue of highly controversial assumptions about that behavior. Without commenting on the merits of those assumptions it is clear that the hypothesis of what operant theorists call instrumental conditioning has enhanced considerably our predictive and manipulative powers with respect to the behavior of children, schizophrenics, and normal individuals in laboratory settings. The leading principle of operant conditioning was first advanced by E. L. Thorndike in the late nineteenth century, and was called "the law of effect."[12] According to this law, and its subsequent amplifications, a positive reinforcement will increase the probability (or the frequency or the intensity) of the recurrence of the kind of behavior which it follows; its omission will decrease the probability (or the frequency or intensity) of the behavior; the elimination of a negative reinforcer will increase the probability (or intensity or frequency) of the behavior emitted on the occasion of elimination; and a punisher will decrease the probability (or intensity or frequency) of the kind of behavior which it follows. This law is highly generic in that it does not specify the units of the behavior which it purports to explain; nor can we determine without considerable further detailed research whether in individual cases what is explained is the probability of a kind of behavior, or the relative intensity of a particular incident of behavior, or the frequency of emission per unit time; nor does the law specify the delay times between events; most important, in applying it to the explanation of particular cases we must provide a specific characterization of reinforcement and punishment independent of the law of effect. Similarly, if this general claim is to have any chance of the sort of theoretical entrenchment a nomological statement bears, reinforcement and punishment will have to be given general specifications independent of the law of effect—lawlike correlations to natural kinds of states, conditions, and events which characterize the organisms in question and their relations to the environment. In other words, to be acceptable as a law that governs the behavior of any organism, human or otherwise, the law of effect will have to pass the very test which our candidate L, of Chapter 5, failed. The demand that it do so is tantamount to the requirement, imposed by operant psychologists upon themselves with as much force as their opponents impose it, that the law of effect, in its particular versions, be testable. These opponents are constantly chiding the proponents of the law with the complaint that the only general characterizations available for reinforcer and punisher make the law of effect an empty tautology. This criticism has some warrant in texts that define "reinforcer" as "any stimulus which, if presented (or withdrawn) contingent on an operant, increases (decreases) the probability of the occurrence of that operant." Such a definition does trivialize the law of effect, but examination of actual practice among psychologists reveals that when they employ the law, they do take steps to specify reinforcers independently. In fact, psychologists do not appeal to the law of effect in its generic form at all in their actual research, even though it bulks large in the textbook expositions of this work. They are interested in formulating and testing a law of effect for, say, pigeons, key-pecking, and food pellets, with a lag time of exactly 0.5 seconds, and a reinforcement schedule of a highly specified type; or again, the law of effect for rats, maze-running, and electric shocks; or monkeys, problem-solving, and free-play opportunities. Reinforcers, the behavior emitted, the rate of reinforcement—all these are specified independently of the law of effect for each of the experiments undertaken. Employing species-specific versions of the law, psychologists have been able to produce replicable experiments confirming generalizations about the constancy of lag times between reinforcements and behavior, rates of response to given reinforcers, rates of extinction of behavior in the absence of reinforcements, degrees of discrimination abilities generated by conditioning, and variations of rates of learning for different schedules of reinforcement. These generalizations have enabled psychologists to shape, control, and predict the occurrences of particular kinds of behavior for a wide variety of species, including *Homo sapiens*, at a level of accuracy that far exceeds the power of any available alternative hypotheses.

These successes have certainly suggested to operant psychologists and to those who employ their findings that the law of effect, at least in its specific forms, provides a far better guess as to (one of) the lowest-level nomological generalization governing human behavior than might any principle couched in terms of the variables of belief and desire. Such theorists offer as the narrowest natural kind into which human behavior falls the operant, and as the narrowest kinds into which its determinant falls, the reinforcer and the punisher. Their candidates have a clear advantage over reasons and actions if only because their cross-species applicability seems to guarantee their purely qualitative character, and thus does not exclude the possibility of laws formulated in their terms on formal grounds alone. Furthermore, the assumption that some version of the law of effect operates at the level of human behavior is strengthened by its relation to at least some of the true singular statements we make about particular actions and their causes in our beliefs and desires. The law of effect will explain why a particular state of belief and desire resulted in a given action on the assumption that the particular states and action are jointly examples of a kind of operant which has been previously reinforced. Naturally, this kind is not to be characterized in the terms which describe the belief, desire, and action in question as such, nor will subsequent, similarly described co-occurrences of beliefs, desires, and actions of the same or other individuals instantiate the same kind of operant manifested on this occasion, although they too might be explained by the law of effect as an instance of another previously reinforced operant. If all those agents whose states instantiate L do so because the agents satisfy a vast, disorganized, and heterogeneous class of variously reinforced operants, then we shall have an even stronger reason to suppose that there are no expressible laws relating reasons and actions, and that L is at best an accidental generalization. Under these circumstances our inabilities to predict actions, given knowledge of reasons, and to control them, given manipulation of reasons, will become clear, for they will reflect the heterogeneity of the schedules, types, latencies, and other variables of reinforcement obscured in the commonality of our ordinary descriptive discriminations of operant states, in terms of beliefs, desires, and actions.

But although these assumptions about the powers of operant theory to explain human behavior give flesh to our own account of the failures of the conventional hypotheses to do so, they do not give any independent ground to believe that this theory actually has such powers. And the fear that it does not, that it may be as much of a dead end as the reason-action explanatory model has proved to be, continues to lurk in the repeated charges that the law of effect is an empty tautology, because reinforcement cannot be given a general enough independent characterization to sustain its explanatory force. Despite the local successes of fitting the behavior of various species to smooth curves of reinforcement for situation- and species-specific reinforcers, the law of effect can function as a nomological statement explaining all of these cases as common reflections of the operation of the same mechanism only if we can find a characterization of reinforcement common to all the species-specific behaviors the theory purports to explain, a characterization independent of the law of effect itself. Failure to do this will make the cost of preserving the narrow operant explanations of the behavior of particular organisms in particular experimental and natural settings nothing less than the surrender of the law of effect as the unifying general law behind them all. In other words, without such a specification, the experimental findings will turn out to have the same status as the true singular causal statements of commonsense psychology: true in their reference, but indeterminate or false in their causal attributions. The actual direction of considerable research in experimental psychology reflects this concern. Once psychologists satisfy themselves that a piece of behavior is emitted because of reinforcement of other instances of its kind, they begin to ask why the particular item identified as the reinforcer has the demonstrated effect; psychologists, like other scientists, want to know what the mechanism beneath their lowest-level generalizations is. If they cannot find one, then the status of their generalizations is called into question. At present, it appears that no uniform mechanism accounts for reinforcement across the range of species and settings to which it has been applied. The earliest and most obvious mechanism hypothesized was that all reinforcers satisfy a biological need of one sort or another. Yet, laboratory animals can be trained by administration of saccharin as a reinforcement, and this is a substance which moves through the body practically unchanged and serves no biological need normally construed. Naturally, if operant behavior is reinforced by saccharin-feeding, there must be some mechanism to account for this, but it is not that of fulfilling narrowly biological needs. A related but somewhat broader theory of reinforcement hypothesizes that reinforcers all reduce tension of some sort in organisms. It has been demonstrated, however, that animals can be trained by reinforcers which it is highly implausible to suppose reduce any tensions. The most widely known of such experiments involves the operant training of monkeys to press a bar by reinforcing the behavior with the opening of a window on their cages that allows them to watch activity in the laboratory. It is hard to say what tensions are thereby reduced. More strikingly, monkeys can be reinforced for puzzle-solving behavior with the solution of the puzzle as sole reward. In an experiment which might be supposed to get at the neurological basis of all reinforcement, James Olds showed that the electrical stimulation of an area of the rat's midbrain provides it with a most powerful reinforcer. The stimulation is of such intensity and frequency that Olds came to call this area of the brain the pleasure center, and to speculate that all reinforcers are causally connected to the excitation of these regions.[13] This sort of result is just what the physicalist would expect, and insofar as it might be supposed that operant theory applies to all overt, nonreflex behavior of an organism, neurophysiological correlation is the only sort of independent specification potentially available to provide independent specifications for reinforcers, as it is in the case of the states of agents that figure in L. A general expectation, however, is a far cry from a practically useful independent specification. Moreover, the same neurophysiological structural complexities and functional redundancies that bedevil attempts to specify beliefs and desires in terms of brain states provide obstacles to the provision of neurophysiological correlates for reinforcers as well. The obstacles are not so formidable for the behavioral psychologist, however, because his typology of operant and reinforcement is clearly applicable beyond the species *Homo sapiens*, and therefore enables him to construct experiments on simpler systems with evidential bearing on hypotheses about more complex ones. The crucial point is that such experiments must ultimately eventuate in manageably small neurophysiological correlates for reasonably broad categories of reinforcers, or they too will turn out to be nonnatural kinds as systematically superfluous as reasons and actions. Pending the successful outcome of such experiments, the law of effect remains at best a candidate for the lowest-level entrenchable nomological generalization governing human behavior.

A different sort of research program in psychology which hopes to provide systematic theories of human behavior and the human cognition resulting in human behavior is that associated with investigations of artificial intelligence and the computer simulation of human behavior. Here the strategy, very roughly, is to specify a set of circumstances in which a human being will produce a certain action, and then to write a program for a machine which, given the circumstances as inputs, will generate the description of the action as an output. For instance, faced with a problem in chess, the human being will decide upon and make a certain move. If a computer can be programmed to generate the same move in the same chess problem, then, it is claimed, there are some circumstances under which the program on which the computer operates serves as the explanation of why, under the circumstances, the human performed the action to be explained.

A distinction is often made between work in artificial intelligence and computer simulation. In the former arena, the aim is to design machines and, more usually, programs that will enable computers to perform activities performed by humans, especially activities which are laborious, routine, and highly time-consuming, although perhaps requiring a great deal of simple calculation, correlation, or other sorts of "number-crunching." Thus, it may be the aim of research in artificial intelligence to design a program that will permit a machine to quickly and accurately complete a quantity of accounting or inventory-taking or multiple regressions or mail-sorting that a human could do, but only in a long period of time, and with the likelihood of a large number of errors produced by boredom, fatigue, inattention, and the like. In writing a program for a machine to take over such functions, we obviously focus on means and powers of the machine that humans do not have; and typically, the program enables the machine to complete tasks by using methods utterly different from those which humans use. Artificial-intelligence research often aims to produce machines that can perform tasks which require from humans the exercise of intelligence, but which can be performed by machines as quickly or more quickly, or more accurately, or more cheaply, by means different from those we use, and which individually need not be described as the exercise of anything like human intelligence. But besides routine jobs that require intelligence of humans but can be performed by computers employing processes that reflect no "intelligence," we may also attempt to program a computer to play chess, a highly nonroutine activity. An artificial intelligence program for a chess-playing machine will usually trade on its brute force powers to apply a few very simple rules to search out the best move to a breadth and depth of alternative move and countermove far beyond human powers of memory and imagination. This, of course, is not the way we play chess: we apply a large number of very complex rules of choice to a much smaller and narrower range of alternative move and countermove possibilities. The differences in strategy obviously reflect differences in strength and weakness between humans and computers. Computers can do simpler calculations faster and remember them; we can do more complex calculations than we have yet been able to articulate and to program; or perhaps, we do not play chess by calculation at all, but by the employment of pattern-recognition capacities beyond the powers of current computers. It cannot yet be said whether artificial-intelligence research will eventuate in a computer with the chess powers of the best human players.

Unlike artificial intelligence, work in computer simulation has a more specific aim than merely building machines that can produce the same "output" as humans, for a given "input." In studies of computer simulation the aim is to produce programs with inputs and outputs which are the same as in human cases and in which the program reflects the "same" processes that we employ to generate the given outputs. Although simulation of human behavior has been judged successful in certain areas, most aspects of this behavior have remained recalcitrant to computer modeling. Attempts to simulate human behavior, however, raise the questions of just what constitutes successful simulation and what sort of understanding of the simulated behavior does such simulation provide. It seems clear that successful simulation of at least some sorts of human behavior need not duplicate all or even most aspects of that behavior. Thus, a successful simulation of chess-playing behavior need not be one in which the machine actually moves the pieces on the board; it would be sufficient for the machine simply to print out a directive to move a specified piece to a particular location, the same piece to the same place that the human, whose behavior is simulated, would move it. It would be a constraint on success that the data fed into the computer be the same as that on the basis of which the agent acts, including the position of the pieces, information about the opponent, the stakes; another important constraint is the requirement that the time scale for easy and difficult moves be at least proportionate between computer and human. Conditions on successful simulation are hard to state in general because they will vary from behavior to behavior and over the degree of simulation possible with given levels of technology.

A question of greater significance is what the simulator hopes to gain by his simulation. The short answer is that the successful simulation of behavior is tantamount to an explanation of it, or at least provides all the resources we require in order to explain it. For while the human represents a black box to the simulator, the machine is, so to speak, a transparent box. Everything about its operation is as well understood as any purely physical system's behavior is: we are acquainted with the laws of physics that govern its behavior, and we know with what program it has been programmed. This knowledge enables us to explain how the machine performs the activities which constitute the simulation and, it is argued, constitute an implicit explanation of the behavior simulated as well. Of course, the conclusion here is a *non sequitur*, as reflection on results in artificial intelligence reveals. For mere simulation may be a case of artificial intelligence, in which the output of the machine is equivalent to the behavior modeled, but the processes employed to produce it are utterly unlike those humans actually use or in some cases could use. The question therefore arises, Under what conditions will a simulation serve as an explanation of or a guide to the explanation of the behavior simulated? This question has received some philosophical treatment. The answers provided by at least one influential writer are worthy of our attention because the results of this chapter turn out to suggest that highly plausible conditions of the sort he advocates, and that many would take to be indubitably obvious necessary constraints on successful simulation, may in fact be serious obstacles to its success.

In *Psychological Explanation* J. A. Fodor argues that in order for a machine simulation to provide an adequate explanation of the behavior of an organism, it must satisfy at least two conditions: (1) the machine and the organism must be weakly equivalent, that is, the behavioral repertoire of the machine must be identical with the behavioral repertoire of the organism; (2) the machine and the organism must be strongly equivalent, that is, the processes upon which the behavior of the machine is contingent must be of the same type as the processes upon which the behavior of the organism is contingent. Fodor writes: "For an adequate simulation to be an adequate explanation it must be the case both that the behaviors available to the machine correspond to the behaviors available to the organism *and* that the processes whereby the machine produces behavior simulate the processes whereby the organism does."[14] As Fodor notes, the requirement of weak equivalence is simply a variant on a standardly imposed adequacy condition for scientific theories. An adequate theory cannot merely be compatible with the known data, for, as noted in connection with Becker's economic theory of behavior, there are an indefinite number of theories compatible with a finite amount of data; an adequate theory must also generate relevant counterfactuals about its subject's behavior. Similarly, a machine must not merely duplicate the actual behavior of an organism in order to simulate it adequately; it must be able to duplicate what the organism *could* do—that is, counterfactuals true of the organism must be true of the machine or be paired with parallel counterfactuals true of the machine. This is the force of the requirement of weak equivalence: that behavioral *repertoires*, and not just actual behavior manifested by machine and organism, be qualitatively the same under some relevant descriptions. This requirement can, of course, be satisfied by a machine that evinces only artificial intelligence, and does not simulate any actual behavior or repertoire of behaviors. For simulation, it is required in addition that there be a functional equivalence between the internal processes of the machine and the internal processes of the organism that are causally responsible for their respective, equivalent repertoires. Thus, if the program for a simulator gives the states of its information-processing in a flow chart or a machine table, then there must be a one-to-one correspondence between the stages described in the program's flow chart and the cognitive stages whereby the organism generated its output.

Since we can produce an indefinitely large number of different programs for any given repertoire, this requirement in effect makes our guide to the discovery of explanatory simulations reports *by the organism* of its own internal states as it generates the behavior simulated, for at present we have no other source of information about the stages which constrain our program choice. Thus, for instance, when simulators attempted to model problem-solving in logic or geometry, they monitored the verbal reports of subjects about the stages in their problem-solving and the reasons which led them to make certain decisions about how to proceed. Then they wrote a program for a computer that enabled it to solve problems of the same kind in similar times, and inferred that the program constituted at least the sketch of an explanation of the human behavior simulated and that, in this regard at least, human behavior is explicable in terms of generalizations drawn from information science.[15] Unfortunately, success in simulation has been limited to a narrow range of behavior, and to a fairly simple level of complexity as well, a range far narrower and simpler than that to which the relatively unconstrained methods of artificial intelligence have been applicable. The reason is in part that the requirement of strong equivalence has limited the alternative programs to those reflecting the *kinds* that humans employ to characterize the factors in their own internal decision processes—beliefs that some propositions are true and desires that others come true. If beliefs and desires are not natural kinds, and computer simulation is constrained to find functional equivalents for them which will generate the same behavioral repertoires in computers as in the subjects simulated, then such simulations are bound to fail. For the absence of laws systematically linking reasons and actions precludes the existence of a behavioral repertoire generated by a causally homogeneous set of internal states of belief and desire. Accordingly, any attempt to generate a repertoire that characterizes its members as examples of a *type* of *action* is bound to fail, since that repertoire is not produced in the human case through the exemplification of a small number of natural kinds by the inner states of the organism. We know that our descriptions of the internal states of a computer are couched in natural-kind terms because such descriptions figure in the physical laws that are employed to build computers, and that are confirmed in their operation. It is because we already have a well-confirmed theory to explain the interrelation of such states and their consequences for the output of the machine that simulation is an appealing strategy for explaining hitherto unexplained phenomena like human behavior. But the criteria of weak and strong simulation in effect oblige us to abandon this advantage by requiring us to characterize the repertoire of the machine in terms drawn from the everyday description of behavior as actions of various sorts, and to limit our choice of programs to ones which are in effect functional equivalents of L and its substitution instances.

If human behavioral repertoires are individuated in terms of types of actions that they exemplify, then the requirement of identity of repertoires condemns computer simulation to the impossible task of finding a causally homogeneous explanation for a heterogeneous class of effects. And if the determinants of these classes are assumed to be beliefs and desires, the requirement of strong equivalence, of identity of processes, condemns computer simulation to search for uniform relations between causally heterogeneous internal states of computers. The upshot is that so long as we retain the commonsense view about the nature of human behavior and its causes, the recalcitrance of most of it to simulation will reflect the imposition of these two apparently unexceptionable constraints. In the attachment of simulators to the common assumption about human behavior and to these adequacy conditions on simulation, we have a far more plausible explanation for the tremendous difficulty involved in providing computer simulations than in the metaphysical differences in kind between men and machines. Of course we need not surrender our adequacy conditions if instead we give up the common assumption that behavior is produced by the joint operation of belief and desire. Naturally, the application of the criteria will turn on the provision of other ways of individuating the behavioral repertoires of humans and the internal determinants of the members of these repertoires. But the descriptive language already available to computer science may provide these means. Since we know how to individuate internal states and outputs of computers in a terminology that does reflect natural kinds—the natural kinds of computer science—it follows from the *assumption* that we can simulate human behavior, that their repertoires and the internal states of humans must be describable in terminology drawn from the language of computer science.

In other words, if we erect as an adequacy criterion for *descriptions* of human behavior and its causes the requirement that these descriptions employ kind-terminology already entrenched in the laws of computer science, then the events, states, and conditions thus described will stand a chance of subsumption under natural laws which on our diagnosis they cannot now do. In effect, this is to turn the adequacy conditions upside down, treating them as constraints on the description of human internal states and behavior instead of constraints on the description of computer behavior and its determinants. Under this assumption, which preserves the plausibility of the requirements of strong and weak equivalence while freeing them from the dead weight of a fruitless descriptive typology, the recalcitrance of human behavior to computer simulation will clearly be seen to reflect empirical, instead of conceptual, obstacles to this sort of explanation of human behavior. Thus, suppose a machine were built that provided a perfect simulation of human behavior—that is, a robot *all* of whose behavior was indistinguishable from that of a human's by some such test as Turing's. Or even better, suppose a robot were built with the same physiognomy and appearance as well as behavior of a human being, and was in fact not detected by anyone but his designers to be a robot. In such a case the designers might think themselves to have in their design for the hardware and in their program at least the next best thing to an explanation of the behavior simulated. Given our adequacy conditions, it would follow from this perfect simulation that humans did behave in accordance with the program of this robot and that their internal and external states were properly described and correctly explained by appeal to the language and directives of this program. Of course, this conclusion will not follow unless *all* human behavior is duplicated. For unless complete duplication were attained, the range of behavior unduplicated would allow the possibility of differences in repertoire that could be explained by attributing different internal processes and programs to the man and the machine.

The upshot of recognizing the nonqualitative, spatiotemporally restricted character of the terms in which human cognitive states are described is that computer simulators should be encouraged simply to maximize identity of repertoires between man and machines, where these repertoires are to be described in the language adapted to the description of computer behavior, without concerning themselves with the question of whether the programs they hit upon reflect any introspective data currently in hand about human cognitive function. And the reason for this is that what is in hand is pretty useless; and successful simulation suggests that the machine program which generates it represents the best evidence available for the actual character of the cognitive states behind the behavior successfully simulated. The degree to which we can satisfy the requirement of weak equivalence is our only measure of the extent to which we are also satisfying the requirement of strong equivalence in machine simulation.

I have argued that desires, beliefs, and actions are not natural kinds, in the last chapter on the grounds that this claim represents the conclusion of an inference to the best explanation of the failure to discover any laws of human behavior, and in this chapter, independently, by appeal to empirical considerations reflected in biological theory. In the light of this conclusion I have examined two research programs for the explanation of human behavior, both to assess their prospects and to illustrate the significance of the reasoning behind my claims about reasons, actions, and natural kinds. In the next chapter the consequences of these claims for the social sciences in general (and not just for these two branches of them) will be examined in more detail, but we may conclude this chapter by turning to a philosophical controversy of the most basic sort, one already touched on at the outset of this chapter. Our conclusions, it turns out, lend new weight to one side of the dispute.

The issue is one often broached in connection with claims like the ones endorsed immediately above about the prospects for simulating human behavior by computers, and the conclusions about mentality drawn from such prospects. It seems an indubitable fact that mental states have content; in the case of desires and beliefs, they have propositional content. We believe that . . ., where a proposition can fill the ellipses, and we desire that some state of affairs be true. States like belief and desire are accordingly called propositional attitudes, and they are described as reflecting the property of intensionality. Now one test of intensionality is that statements attributing intentional properties to particular items, like psychological states to human beings, are incapable of absorption into the extensional logical apparatus that seems to suffice for analyzing and regimenting the statements and inferences of mathematics and the natural sciences. A simple illustration of this recalcitrance of intentional discourse is provided by the following seemingly valid inference from apparently true premises to a presumably false conclusion:

Lady Astor desired to sail on the largest ship afloat in 1912.
The largest ship afloat in 1912 was identical to the ship that struck an iceberg, sank, and caused Lady Astor's death.
Therefore,
Lady Astor desired to sail on the ship that struck an iceberg, sank, and caused her death.

Since the conclusion is false, and the inference-form unexceptional, the premises must be ill-formed for purposes of logical manipulation. The trouble is usually diagnosed by noting the intensional character of the first (and sometimes also the second) premise, and such premises reflecting propositional attitudes are excluded from theories which, like those of physics, are regimented in accordance with extensional logic. Now, if computers, or brains, for example, are purely physical systems, and no physical description of their properties and states will generate invalid inferences of the sort illustrated, then they and all their activities can be completely described in propositions that make no recourse to intensionality. In consequence, since psychological states do seem to manifest intensionality, the attribution of such states to computers (which the simulator hopes to make) or to brains (which the neurophysiologist hopes to make) will require an analysis of intentionality that enables us either to attribute it to computers or brains or to translate intensional statements into ones that are not intensional, that do not generate invalid inferences of the form illustrated. Now the first alternative is counterproductive of systematic regimentation of scientific theories, for it would extend the range of theories not amenable to treatment by the resources of extensional logic. But the second seems unlikely of fulfillment. None of the large number of attempts to analyze belief, desire, or any of the other propositional attitudes has eliminated their intensional features, although some have translated one sort of intensional feature into another. Indeed, it has been a widely held philosophical thesis associated, since the nineteenth century, with the name Brentano that intentionality is an essential feature of psychological attitudes, and that therefore any science of such attitudes, like psychology and the other social sciences, must as a matter of logic be autonomous and irreducible to the natural, extensional sciences.

But this conclusion, frequently cited to insure the insulation and irreducibility of mentalistic to physicalistic language, is a double-edged sword. Although no physicalistically motivated account of mental states and events has successfully analyzed away the intentional element of such states and events, the antiphysicalist has been able to do no more than to assert that mental phenomena have this property of intentionality. They have been equally unable to explain why it is that intentionality is limited to this range of phenomena or what this intentionality consists in. It is no answer to either of these questions merely to say that propositional attitudes are intentional because they are conscious states of agents or are related to such states, or that they are intentional because they are psychological or mental, nor will it do to explain their intentionality in terms of their intensionality, in the invalidity of otherwise unexceptionable arguments in which they might figure. The first answer is unsatisfactory because the question, Why are psychological or mental states intentional? is the same question about the intentionality of propositional attitudes all over again. The second answer is similarly unavailing. In asking what intentionality consists in, we want to know why it is intensional, why it generates invalidity in the way it does, not merely that it does generate such anomalies for logic. Of course, the explanation of intentionality or intensionality may not excite the interests of those who cite it to establish the autonomy of the sciences of intentional phenomena. But the existence of such an explanatory dead end for features of so complex a system as a human being is plainly intolerable for the empiricist. Thus, W.V.O. Quine writes, "One may accept the Brentano thesis either as showing the indispensability of intentional idioms and the importance of an autonomous science of intention, or as showing the baselessness of intentional idioms and the emptiness of a science of intention. My attitude, unlike Brentano's, is the second. To accept intentionality at face value is . . . to postulate translation relations as somehow objectively valid though indeterminate in principle relative to the totality of speech dispositions. Such postulation promises little gain in scientific insight if there is no better ground for it than that the supposed translation relations are presupposed by the vernacular of semantics and intention." Quine's argument here is similar to that advanced above in connection with the adequacy conditions of computer simulation, and in fact is but a restriction of that argument from behavior in general to linguistic behavior in particular. Requiring computers to satisfy intentional descriptions which are themselves inexplicable promises no more gain in scientific insight than requiring speakers to satisfy the same intentional properties when they decode each other's linguistic signals. Because of the scientific sterility of intentional notions, Quine continues, "if we are limning the true and ultimate structure of reality, the canonical scheme for us is the austere scheme that knows no . . . propositional attitudes but only the physical constitution and behavior of organisms. . . . If we are venturing to formulate the fundamental laws of a branch of science, however tentative, this austere idiom is again likely to be the one that suits."[16]

How is this dispute to be settled? Brentano and his party have the certainty that comes with direct introspective access to the existence of these intentionally characterized states. Quine and his cohorts have the admission of the inexplicability and therefore systematic sterility of intentionality, as well as its recalcitrance to regimentation in a logic that seems to suffice for all other scientific purposes. Is the argument a standoff? Do the conclusions of this chapter—that for considerations independent of issues controversial between the two parties, propositional attitudes are not natural kinds and cannot figure in laws—tilt the balance in Quine's direction? I should say the answer is yes. For they show, minimally, that Brentano's science of intention will be a science without laws, and therefore no science at all, strictly so called. More arguably, these conclusions and their applications to behaviorism and computer simulation suggest that there may well be natural kinds under which each and every particular propositional attitude falls which are clearly not themselves intentional. Most importantly, the considerations of this chapter help the empiricist and the physicalist explain away the phenomenon of intentionality as without systematic scientific significance in the description of human beings and their behavior; and they do so without appealing to philosophical theories controversial between Brentano's exponents and his opponents. The physical irreducibility of propositional attitudes and the intentionalist's own inability to explain why psychological states are intentional both hinge on the fact that such states when intentionally characterized, involve tacit reference to a spatiotemporal particular (the species *Homo sapiens*) and so cannot figure in synthetic general statements of a lawlike kind. But if the explanation of the apparently general fact that all psychological states or propositional attitudes are intentional requires the citation of other general laws from which the explanandum is derivable, then plainly no such explanation will ever be available. Singular statements cannot be derived from general ones. The behaviorist's persistent inability to analyze intentional states into statements about their extensionally characterized behavioral consequences is explained by the fact that intentional description of the behavior's causes precludes the expression of nonaccidental general statements relating these causes and their effects. For these causes are not characterizable in purely qualitative predicates. Indeed, on our hypothesis, the discovery of a true synthetic statement relating an intentionally characterized psychological *kind* of state to a nonintensionally characterized kind of behavior would be nothing short of a miracle. This is why behavioral analyses of such concepts as "belief that . . ." or "desire that . . .," or more specifically, "belief that it is raining on 31 August 1996 in the center of Salzburg, Austria" are invariably either open to obvious counterexample, or closed to empirical overthrow altogether. This failure, paralleled in the similar impossibility of giving a true general neurophysiological characterization of psychological states, is what their intentionality consists in. Because we cannot independently characterize them in consequence of the systematically undescribable connection which each and every psychological state has to some particular brain state and to some particular behavior (within or at the periphery of the body), we are reduced to characterizing them in terms of independent states of affairs (described in the propositions they "contain") with which the intentional states bear no invariable causal connections whatever. Because they bear no such relations, they generate invalid arguments, are irreducibly nonphysical and nonbehavioral, and their attribution to anything whatever is utterly inexplicable. All of this follows from the arguments of this chapter, and suggests, with Quine, not the autonomy of a science composed of such statements, but its impossibility.

7

Human Science and Biological Science

If beliefs, desires, and actions are not the natural kinds in which we can expect to classify the causes and effects which social science seeks to explain, what are the narrowest natural kinds under which we can with confidence and with practical usefulness subsume the explananda of these subjects? Are there in fact any natural kinds already known to us in which we can couch true and useful general statements about human behavior? If, as noted in Chapter 6, the lowest-level laws governing human beings were those of a narrow branch of organic chemistry, so that the narrowest natural kinds in which we and our behavior figured were given by the descriptions and predicates of this branch of organic chemistry, then we could neither explain any particular human behavior beyond the most schematic level, nor could we predict or control that behavior by any means likely to be at our disposal in the foreseeable future. For no matter how well acquainted we might be with the laws of this branch of organic chemistry, we would be unable to ascertain the satisfaction of explanatory or predictive initial conditions which together with the laws might imply the explanandum event, even if we could describe this event in terms precise enough to subsume it under the laws. And similarly, any attempt to employ these laws to control behavior would be thwarted by a parallel inability to satisfy the boundary conditions required by the laws. If this state of affairs were the case, it would be reasonable to conclude that a *science* of human behavior in the normal understanding of that word is, as a matter of practical interest, *impossible*. Under such circumstances, the pretensions of sociologists or economists, psychologists or political scientists, to provide complete or completable explanations of particular events (as opposed to merely retrospective true singular statements about their causes, unjustified by any sustaining regularity) would be empty; and the expectations of policymakers, politicians, propagandists, and psychotherapists that *reliable* predictions and technologies of control are forthcoming from these sciences would be equally vain. We would have to either dismantle or find new rationalizations for academic departments, institutions of research, charitable foundations, professional schools, learned journals, and all the other impedimenta of our assumption that there could be sciences of human behavior.

Of course there are those ready to insist that they are acquainted with a range of natural kinds that answer to our need for nomologically permissible and practically useful description of human behavior and its determinants: operant and classical conditioning and computer simulation, as we saw in Chapter 6, are two coherent candidates for theories generating the required practicable kinds with which to classify human behavior. And there are other candidates ready to hand. Thus, Freud's theory of the unconscious causation of behavior, both normal and abnormal, implicitly represents an alternative typology to that of common sense for the description of action and its determinants. Insofar as Freud's theory represents an attempt to extend the sorts of determinants mentioned in a statement like L to cover states, conditions, and events described as nonconscious or unintentional, it constitutes an implicit redefinition of the terms of L, and thus a new typology for the causes of human behavior. Of course, as Freud originally viewed the theory, its causal determinants were surrogate descriptions for the neurophysiological causes of behavior. As such, Freud's theory, like Skinner's, awaits the provision of independent neurophysiological specifications for its variables. Until these are provided, the theory remains unentrenched in a network that could give it general explanatory or predictive power. Of course, quite independently of such specifications, we already know that the theory is much less reliable, even as an accidental generalization, than Skinner's is. Severed from its neurophysiological interpretation, as it has been by Freud's followers, it has all the weaknesses of a general statement like L: that insofar as its terms are intentional, they are not purely qualitative, and that in any case the theory has not even been stated in an agreed-upon form by its own current exponents. Freud's theory and its vagaries are paradigmatic: each of the seemingly endless succession of new speculations about human behavior broached in learned or popular forums manifests one or another of the features that vitiate Freud's theory as a body of nomological generalizations practically useful in the explanation and prediction of behavior.

None of the intentional kinds offered in anthropology, sociology, political science, economics, or psychology (clinical or experimental) justifies any confidence. On the other hand, none of the natural kinds that our confidence in physics or chemistry recommends to us is of any practical moment for the explanation of human behavior. Yet, just as considerations from biological theory were crucial in the explanation of why the kind terminology of the social sciences cannot be natural, so considerations from that same subject direct us to the narrowest natural kinds that we can be *sure* subsume human behavior and provide us with a clear guide to the extent that these kinds will provide explanation and prediction. That is, we can be as sure that the kinds of biological theory are *natural* kinds as we are confident in the truth of the lawlike statements of biology. Since our confidence in at least some biological laws is very great, and certainly far greater than any reasonable confidence in putative psychological laws, for instance, we are committed to treating human beings and their states as falling under the kinds presupposed by these laws. But the laws of biology are clearly the lowest-level laws *with which we are currently acquainted*. Biological laws are derivable schematically or in detail from the laws of physics, organic chemistry, and biochemistry, while the laws of these latter theories are not derivable from them. Biological laws are about a range of entities whose components we believe to behave in accordance with the laws of chemistry and physics. It follows, therefore, that the narrowest natural kinds in which we can at present expect humans and their behavior to fall are the natural kinds of biology. The fact that it is incompatibility with biological laws or their conceptual presuppositions about species which vitiates the candidates offered by social sciences is further evidence for this conclusion. Finally there is the obvious fact that human beings are living systems, and therefore paradigm subjects of biology. This last consideration suggests the general relevance of biology to the explanation of human behavior, a relevance that no one ever doubted. The other considerations suggest more than relevance, however; they indicate that if our desire is to formulate practically significant general statements that we can be confident of, given current knowledge, the kind-terms in those general statements must be those of biology, and not of psychology, economics, anthropology, sociology, political science, and so on. Moreover, these considerations suggest that we can expect no more and no less completeness of explanation and accuracy of prediction than is in principle attainable in biology at present or in the future.

This conclusion, although controversial and significant for the future study of human behavior, is as yet much too general to provide firm or specific expectations. The conclusion justifies the employment of empirical methods in the study of human behavior to the extent that such methods are presumably justified in biology. It does not undercut our confidence in the singular truths of everyday life about particular human events and their causes; but it also leaves these statements untouched. Are they somehow to be systematized as data, or ignored as anecdotes? The conclusion assures us that there are laws governing human behavior and tells us that they are as practically useful as the laws of biology, for they are the laws of biology. But it leaves unanswered the question of precisely how useful these laws are, given our actual purposes, and therefore leaves unanswered the question of whether these purposes are reasonable ones. In this chapter I shall attempt to expand on the suggestion that a commitment to empiricism justifies the appeal to biological theory and in the next chapter to answer these questions. We shall be able to assess their significance for the positive proposals which must go hand in hand with the empiricist's diagnosis of the failures of conventional social science.

Of course, the application of biological modes of thought to questions hitherto addressed in conventional social science is not new. Almost immediately after the reception of Darwin's theory of natural selection, philosophers and social thinkers began applying its claims to the explanation of social phenomena. Most notorious among these expositors of Darwin was Herbert Spencer.[1] Social Darwinists have recurrently appeared in the history of twentieth-century social science without leaving much of a permanent impression. Perhaps the only permanent contribution which biological theory has made in the social sciences is its explanation of apparently purposeful goal-directed phenomena like evolution in terms of the operation of completely nonpurposive forces like variation and selection. Social theorists hoping to explain the persistence of a social institution in the absence of any intention on the part of individuals to see it preserved have been able to appeal to a mechanism of natural selection, together with the hypothesis that the institution constitutes an adaptive feature of the society. As philosophers of social science have been quick to point out, this sort of explanation cannot account for the origin of such institutions, nor can it be subject to any objective test, interconnected with other findings and theories, employed to predict continued persistence of this or any other item, unless the character of its comparatively greater adaptiveness is made clear independent of its persistence.[2] This is a demand which functionalist theorists in the social sciences have yet to satisfy, and their failure to do so has severely limited the influence of biological theory among conventional social scientists. In recent years, however, the application of biological modes of thinking to the explanation of phenomena previously within the exclusive ambit of social science has substantially increased, and the source of this activity has not been within the conventional social sciences, but has centered in a distinctively biological network of researchers. *Sociobiology* represents the self-conscious extension by *biologists* of their work to explanatory and predictive problems in the social sciences, and not its importation into the traditional subjects of social science by *social scientists*. Naturally, therefore, the biological sophistication of this work has been much greater, while the attention to traditional problems and strategies in social science, not to mention the sensibilities of social scientists, has been much less than that previously evinced by exponents of the application of biological concepts and theories to human behavior.

Quite independently of any careful examination of the specific aims and claims of sociobiology with respect to the explanation of human behavior, the body of theories that go under this rubric is often rejected by social scientists on the grounds that their exclusive dependence on Darwinian theory reflects an obviously false supposition called the assumption of *biological determinism*. This assumption is usually expressed as the claim that all human behavior, no matter how individuated and no matter how apparently different in its extent or scale, is the causal product of hereditary factors exclusively. Thus, sociobiology is sometimes rejected because its detractors attribute to it the claim that the sufficient cause of every particular human action, from Jefferson's drafting of the American Declaration of Independence to the Pope's latest blessing in Saint Peter's, and the cause of every nonintentional, merely autonomic bit of reflex behavior as well, is to be found in the human genome. Since this claim is patently false, those who see it as an essential doctrine of sociobiology have no difficulty concluding that the latter's pretensions to explaining human behavior need not detain them. A weaker version of the thesis of biological determinism sometimes attributed to sociobiology is also considered sufficiently absurd as to enable us to ignore the claims of this subject. It holds that sociobiology requires or implies that all of the traits, dispositions, capacities, abilities, and disabilities that we attribute to humans to explain regularities in their behavior are determined exclusively and completely by their genetic inheritance. This thesis, though more plausibly attributed to sociobiological theories, is weaker because it does not entail that genetic inheritance is causally sufficient for each and every particular event whose constitutive object is a human being; rather, it suggests that the entire repertoire of an individual's behavior is narrowly fixed from conception by his genotype, and that whatever behavior is produced from among the limited range of behaviors figuring in this narrow repertoire is determined by any one of so vast, heterogeneous, and common an array of environmental contingencies that these latter, causally necessary conditions hardly bear mention. More specifically, since the different environmental triggers of behavior are so indiscriminate in their elicitation of behavior, and the genetic burden is so narrow in the range of behavior it "permits," embracing this version of biological determinism forces us to focus our theoretical attention almost exclusively on the genetic determinants of the narrow repertoires of behavior. This version of the thesis of biological determinism is often declared to be false on ideological grounds. It is proclaimed to be a racist or sexist or anti-working-class theory, one whose truth is incompatible with the interests of certain social groups, and compatible only with the interests of other, opposing groups. Therefore it is declared either to be a body of false propositions or at any rate a collection of disguised ideological rationalizations. In either case the sociobiological theories with which it is associated are rejected as unscientific.[3]

The proponents of sociobiology can hardly defend themselves against this view by offering a neutral definition of their discipline that does not commit them to any such thesis. Thus, to define sociobiology in the way that E.O.

Wilson, the preeminent exponent of this subject, does, as "the systematic study of the biological basis of all forms of social behavior, in all kinds of organisms, including man,"[4] is no defense against the attribution to them of an allegedly false presupposition of biological determinism. For such a definition does not distinguish their discipline from that of biology *tout court*, nor reflect their belief that the application of biological reasoning to human behavior will provide real insights of the sort that conventional social science has hitherto sought in vain. This sort of definition provides no more guarantee that sociobiology constitutes an interesting or worthwhile endeavor than the definition of a subject like "sociophysics" or "sociochemistry" as "the systematic study of the physical or chemical basis of social behavior." If, as we have argued, human behavior is subsumable under the natural kinds of physics and chemistry, then construed quite abstractly, there certainly are such subjects as sociophysics and sociochemistry. In fact, just as biological concepts have repeatedly been applied literally or metaphorically within the social sciences since Spencer's time and before, so too have there been attempts to literally or metaphorically develop social theories that trade on notions from mechanics or chemical reaction theories. But these have all been failures, and have disappeared after running a course of fashionability reflecting popular interest in the subjects, physics and chemistry, from which their analogies are taken. The reason for these failures is, of course, that the natural kinds of these theories will not enable us to generate nomological generalizations useful in connection with the interests of social science. The existence of a real subject, with useful laws of human behavior, requires more than so banal a definition. In fact it requires something like the assumption of biological determinism which sociobiology's detractors have attributed to it. Only on an assumption at least something like this one will the sustained search for determinants different from the ones conventional social science focuses on be motivated; only on some such assumption can we expect to find in sociobiological research nomological generalizations that are practically useful or that at least clearly delimit the degree of completeness in explanation and precision of prediction we can expect in regard to human behavior.

In fact, sociobiologists like Wilson are committed to something like this weaker thesis of biological determinism, cited by their opponents as a *reductio ad absurdum* of their claims. A careful examination of the views to which they are actually committed, and an analysis of the consequences of these views for the explanation and prediction of human behavior will not only help refute the shrill arguments of sociobiology's detractors, but also help plumb the consequences of the hypothesis that the narrowest natural kinds under which human behavior falls are those provided in biology.

In the *locus classicus* of the subject, *Sociobiology*, Wilson broaches the topic of the causation of behavior, and writes that "the prime movers of evolution [variation, mutation, selection] are the *ultimate* biological causes. . . . The anatomical, physiological and behavioral machinery they create constitute the proximate causation of the functional biologist. Operating within the life spans of organisms, and sometimes even within milliseconds, this machinery carries out the commands of the genes on a time scale so removed from that of ultimate causation that the two processes sometimes seem to be wholly decoupled."[5] This passage sounds so close to the extreme claims of the strong version of biological determinism that it makes the conclusions of the opponents of sociobiology seem quite plausible. Could Wilson literally intend to suggest that the functional machinery of the organism responds in its moment-to-moment states to states of the genome at the time of development in the way that overt movements of soldiers respond to commands of the drill sergeant? It sounds as though particular events at the genetic level are causally sufficient for much later and much more complex particular events at the levels of anatomy, physiology, and behavior. For instance, it appears as if Wilson is committed to the claim that each blink of a particular eye was predetermined at the instant of conception by the establishment of the individual's genetic inheritance. Surely this is false. Genetic inheritance is not even causally sufficient for the existence of any individual eyelid, let alone the particular occasions upon which it blinks. And of course Wilson knows this perfectly well. We cannot attribute this sense to his claims. His point may be the much weaker one that there is a causal connection between genetic inheritance and each and every event involving the anatomical, physiological, and behavioral machinery of the organism. Yet this claim suffers, not from falsity, but from obviousness, and provides no grounds for focusing on, say, hereditary variables in the explanation of behavior.

The force of the claim should really not be treated as one about causation at all, but rather, as one about the prospects for explanation of anatomical, physiological, and behavioral events. For in the next paragraph Wilson writes, "Most psychologists and animal behaviorists trained in the conventional psychology departments of universities are non-evolutionary in their approach. Yet like good scientists everywhere they are always probing for deeper and more general explanations. What they should produce are specific assessments of ultimate causation rooted in population biology. What they typically produce instead are the nebulous independent variables of theoretical psychology—attraction, withdrawal, thresholds, drives, . . . tendencies. And this approach creates confusion, because such notions are ad hoc and can seldom be linked either to neurophysiology or to evolutionary biology, and hence to the remainder of science."[6] So Wilson's point is that the findings and theories of psychologists about animal behavior do not reveal the proximate causes of that behavior at a level of description capable of any sort of systematization or generalization, and that their characterizations of these proximate causes are nebulous and *ad hoc*. They are so, apparently because they are disconnected from the remainder of science. In particular, studies of the proximate causes of animal behavior are unsatisfactory and employ notions that are nebulous and *ad hoc* because they can be linked neither with neurophysiology or evolutionary biology. This will explain why Wilson recommends that animal behaviorists produce specific assessments of ultimate causation rooted in population biology, instead of nebulous discussions of the proximate causation of particular events. Thus, Wilson's view will be explicable provided we are given a reason to suppose that only through a connection with either neurophysiology or population biology can the claims of students of animal behavior be linked to the rest of science. Wilson's argument for this claim is in the form of a speculative extrapolation with an important metaphysical and methodological conclusion:

> Although behavioral biology is traditionally spoken of as if it were a unified subject, it is now emerging as two distinct disciplines centered on neurophysiology and on sociobiology respectively. The conventional wisdom also speaks of ethology, which is the naturalistic study of whole patterns of animal behavior, and its companion enterprise, comparative psychology, as the central unifying fields of behavioral biology. They are not. Both are destined to be cannibalized by neurophysiology and sensory physiology from one end and sociobiology and behavioral ecology from the other. . . . [This] seems to be indicated both by the extrapolation of current events and by consideration of the logical relationship behavioral biology holds with the remainder of science. The future, it seems clear, cannot be with the ad hoc terminology, crude models, and curve fitting that characterize most of contemporary ethology and comparative psychology. Whole patterns of animal behavior will inevitably be explained within the framework, first of integrative neurophysiology, which classifies neurons and reconstructs their circuitry, and second, of sensory physiology, which seeks to characterize the cellular transducers at the molecular level. . . . *To pass from this level and reach the next really distinct discipline, we must travel all the way up to the society and the population.* Not only are the phenomena best described by families of models different from those of cellular and molecular biology, but the explanations become largely evolutionary. There should be nothing surprising in this distinction. It is only a reflection of the larger division that separates the two great domains of evolutionary and functional biology.[7]

In our terms we may express this claim as the view that above the natural kinds of neurophysiology—that is, biochemistry—the *next narrowest class* of natural kinds under which animal behavior falls is the set of kinds that figure in evolutionary theory, population biology, behavioral ecology, and sociobiology. There is no set of natural kinds between those that describe the behavior of small numbers of nerve cells and those that describe the behaviors of large groups of organisms homogeneous with respect to evolutionary forces; there are no natural kinds which can subsume organic material organized only to the level of the individual biological organism and its behavior. Equivalently, there are no discoverable general laws standing partway in generality between the biochemical laws in which neurophysiological regularities are expressed and the evolutionary laws in which the findings of ecology and sociobiology are expressed. If this is so, then the notions, the nebulous independent variables, of theoretical psychology will turn out to be *ad hoc*, nebulous, and incapable of linkage to the rest of science, just as Wilson claims. And the terminology, the models, the curve-fitting, of contemporary ethology and comparative psychology will turn out to have been *ad hoc*, crude, and arbitrary just because the concepts in which they are expressed turn out not to reflect natural kinds. Consequently, short of being condemned to doing neurophysiology, the psychologist has no alternative but to turn his attention to what Wilson calls the ultimate causes of behavior in the organism's genes and the evolutionary consequences of its genetic burden.

If no systematic, practically useful theory is possible in the nomological vacuum between functional biology and evolutionary biology, because the classes of events that we are capable of individuating at those intermediate levels of aggregation are not homogeneous with respect to causes and effects, then we can provide an interpretation for the version of biological determinism to which Wilson is committed, which is neither absurdly false on its face nor so vacuous as to have no influence on scientific research. It is not that the prime movers of evolution are the ultimate biological causes creating the proximate causes of individual instances of behavior. Rather, the prime movers of selection, variation, and mutation provide the most fundamental, the lowest-level systematic causal *explanations* available for individual instances of behavior to any level beyond the hereditarily determined repertoire of which it is an instance. Of course we can make true singular statements about particular human events and their proximate causes, and we can undertake neurophysiological examinations of these particular events as well. But the former sorts of statements will be forever isolated from nomological explanations of their truth, and the latter sorts of investigations will never eventuate in laws that will completely explain the occurrence of the event in question as an instance of the sort of event we initially described. Moreover, the neurophysiology of the organism is a clearly specifiable function of the ultimate causal variables of selection, variation, and mutation. Naturally, neurophysiology will vary from organism to organism in some of its fine structure. But this variation will not be the subject of many useful general laws, in part because it is the product of a highly nonhomogeneous class of environmental forces, whose characterization will usually be species-specific and therefore incapable of nomological expression. These are the very forces whose effects appear to decouple the proximate and the ultimate causal factors in animal behavior. They do not decouple these factors, but they do prevent our giving a nomological explanation of anything more than organisms' repertoires; they preclude the detection of laws at the level of individual human events, as opposed to laws about human dispositions, hereditarily individuated.

Here we have a sense in which the sociobiologist is properly described as committed to a thesis of biological determinism; the closest we can get to a deterministic or indeed any sort of scientific theory that is capable of explaining any part of the behavior of organisms above the level of neurophysiology is the theory of natural selection together with this presupposed genetic theory. Thus, a scientific study of animal behavior should focus on the ultimate causally necessary (but not sufficient) determinants in heredity of this behavior, not because they are any more than causally necessary for it, but because these variables figure in the lowest-level laws there are about animal behavior; they are the narrowest natural kinds that subsume the subject matter of ethology and psychology. This is the view to which Wilson is committed by his claim that there is no autonomous discipline between neurology and sociobiology; its truth is necessary if we are to write off ethology and conventional studies of animal behavior as inevitably *ad hoc*, crude, arbitrary, and nebulous (though not conceptually vitiated). Most important, it is a defensible version of the thesis of biological determinism that sociobiology requires to justify its pretensions to preempt other conventional sorts of explanations of the behavior of organisms, nonhuman and human. Without such a claim sociobiology has no leverage with which to displace its competitors for theoretical acceptability, no call to our attention as more than just another speculative alternative to their theories, findings, and failures. Make no mistake: although Wilson claims only to displace behavioral biology, ethology, and comparative animal psychology, the real significance of his argument is sociobiology's displacement of all the social sciences normally conceived. For economics, political science, sociology, and anthropology are only branches of the ethology and "behavioral biology" of one species: *Homo sapiens*. As such, they are superseded by sociobiology to just the same extent as the subjects that stand between the functional and evolutionary biology of all the other species are superseded by it. The claim that to pass from the level of neurophysiology and reach the next really autonomous discipline, "we must travel all the way up to the society and the population" involves the supersession of the traditional subjects of the social sciences as much as any other theories about organisms couched in language somewhere "between" neurophysiology and population biology.

Thus, the threat many conventional social scientists feel to emanate from sociobiology is justified, insofar as that theory is committed to preempting theirs. And the attribution to sociobiologists of a commitment to some version of biological determinism is not so far from the mark either. They are committed to something we might call *explanatory biological determinism:* the thesis that scientifically acceptable explanations for human behavior are available only at the level of the heritable dispositions which it reflects or the neurological states which underlie it. In fact, insofar as sociobiology is supposed to preempt the traditional social sciences—as opposed merely to being another speculative alternative to conventional theory or a noncompeting, extended biological theory of the behavior of nonhuman social organisms (ants, beavers, baboons, etc.)—it even more strongly requires the thesis of explanatory biological determinism and a convincing argument for this thesis. At present, sociobiological theory has very little by way of firm and detailed explanations or borne-out predictions about human behavior to show for itself, yet it is clear that its proponents construe it as much more than a mere alternative to other intriguing speculations in social theory, and as more than a theory of the social insects and monkeys. In the absence of an argument for preemption based on firm success in findings and explanations, in the resolution of long-standing anomalies, and in technological or practical application, sociobiology can substantiate itself only by appeal to conceptual and methodological considerations of the sort that the thesis of explanatory biological determinism reflects.

This claim of explanatory biological determinism with which I have armed sociobiology has a number of advantages. It is clearly superior to the strong and weak theses of biological determinism. Unlike the strong thesis of biological determinism, this thesis is not obviously false. It shares with the weak thesis the property of explaining why there are as yet no general laws in the social sciences (i.e., between the levels of neurophysiology and evolution) by implying that there are no such laws to be found. But unlike the weak thesis, the explanatory claim is not committed to the claim that all of the traits, dispositions, capacities, abilities, and disabilities that *we attribute* to humans are entirely or largely hereditary in their scope and character. It is not committed to this thesis because explanatory biological determinism is not committed to treating any of the terms in which we describe these traits, dispositions, capacities, abilities, and disabilities as natural-kind terms that will figure as the narrowest natural kinds in which biological laws will be expressed. Explanatory biological determinism is committed to treating as the traits, dispositions, and so on that provide the most specific nomological explanations of human behavior all and only those properties of organisms that biological theory requires us to describe as *phenotypes*. That is, only those properties of organisms whose distribution and transmission obey the laws of heredity, and therefore are natural kinds, will constitute the traits to whose fixity the theory is committed. It is to the explanatory fundamentality of these that explanatory biological determinism commits the sociobiologist. Since most of the traits and dispositions which we currently attribute to human beings are clearly not phenotypes, it follows that unlike the weak thesis, explanatory biological determinism is not committed to their fixity. Moreover, since it is not, it circumvents the claim, true or not, that sociobiology is merely the ideological expression of anti-working-class or racist or sexist interests. For such arguments proceed by attributing to sociobiologists the views that dispositions like intelligence, or capacities like earning power, or traits of sex role, as ordinarily defined, are fixed, and that this fixity determines irrevocably that society must manifest its current character. Burdened with this view, sociobiology is stigmatized as a conservative or indeed reactionary or fascist theory whose falsity will be revealed in the utopian aftermath of a coming social revolution. Attempts to meet this criticism calmly sometimes fall on deaf ears. Even where hearing remains acute, the protestations of innocence and of a desire simply to let the facts take us where they may seem unconvincing. But in the light of the explanatory version of biological determinism, we may see that the issue of whether sociobiology's commitments provide an ideologically charged foundation for the status quo is simply a red herring. For the theory is not committed to hereditary fixity of the current distribution of traits and dispositions that we have come in ordinary life to identify and distinguish from one another. It is committed to the hereditary distribution of traits which satisfies the laws of heredity, and it individuates and distinguishes traits by appeal to whether candidates satisfy those laws. As such sociobiology is committed to the truth of the laws of heredity, and to whatever fixity those laws demand. To attack sociobiology on the grounds that it claims that there are hereditarily fixed traits and dispositions is simply to reject the most central and well-entrenched theoretical discoveries of biology. Opponents of sociobiology may want to do this, but they should not reject sociobiology under the illusion that in doing so they can preserve the rest of science and its methods undisturbed. It may turn out that the phenotypes to whose fixity sociobiology commits us are no more ideologically significant than eye color, or that they permit of so broad a range of repertoires as to be consistent with a vast range of social structures; on the other hand, it may turn out that sociobiology's findings do reveal a real restriction on possible social arrangements. Which of these turns out to be the case is as yet unknown. The point of the doctrine of explanatory biological determinism is that there is no secure explanation or prediction about individual or group behavior at a level of detail below phenotypic regularities. To this extent it undercuts the security of both utopian and anti-utopian claims about general relationships among socially significant variables.

The thesis of biological determinism, understood as one about the limits to nomological explanation of human behavior, is clearly of a kind that sociobiology needs, and among that kind also plainly more reasonable and more plausible than available alternatives. Indeed, so far as plausibility is concerned, all of the considerations which lead one to conclude that the social sciences will never find laws of human behavior are considerations in favor of explanatory biological determinism, and *a fortiori* arguments for giving socio- biology a chance where conventional social science has failed. Whether our thesis is one we can plausibly attribute to Wilson instead of the strong and weak theses between which he seems to waffle is debatable; what is not de- batable is that it is all he needs to express the claims of his methods to pre- empt those of conventional thinking about human behavior. Moreover, so far as our own problem is concerned, the thesis of explanatory biological determinism not only reflects but gives positive content to the empiricist diagnosis of the failure of social science here advanced. It reflects the di- agnosis in its own dependence on the theories of an already established science for its natural kinds, and the thesis gives positive content to that di- agnosis by providing an implicit prescription and prognosis for further scienti- fic study of human behavior. The prescription is that social science should adopt a new typology, one with a better chance of reflecting natural kinds, and so eventuating in nomological generalizations. The prognosis is that con- centration on neurophysiology and population genetics is likely to succeed if any research program is.

Thus, our desires to preserve empiricism, constrained by a universal com- mitment to the truth of some of our singular causal statements about particular human events, and also to an explanation of the failure of social science to discover the laws of human behavior that empiricism demands, appear to lead us to embrace the perspective on human behavior offered by sociobiology. That an unswerving allegiance to empiricism should lead to the erection of biology as a methodological and substantive paradigm for social science is, however, quite surprising. For biological methods and biological findings have traditionally been problematical for an empiricist philosophy of science. For instance, the teleological language used by biology in explain- ing the operation of biological subsystems like organs or processes like photo- synthesis is foreign to the language of the physical sciences. This is a problem for the empiricist, because his commitment to the unity of science obliges him either to show how teleological language can be analyzed into the non- teleological language of physical science, or to read teleological biology out of the natural sciences altogether. If the analysis cannot successfully eliminate teleological language from biological explanations, and if biology cannot be pursued without appeal to such concepts, then its explanations will differ in kind from the mechanistic ones that figure in chemistry and physics, and the empiricist will face a choice between accepting biology as a body of system- atized knowledge or giving up his thesis of the methodological unity of the sciences.[8] More than one philosopher of science has found the rejection of biology as an autonomous natural science inviting, especially since these philosophers' incomplete knowledge of biology leads them to mistake its accidental generalizations for its most well-established general statements.

The physicalist's commitment to the substantive unity of science and to the reducibility of the laws of each science to those of more basic ones, and ultimately to the laws of physics, also causes him to be uncomfortable with the scientific status of biology and with the claim that it should be a paradigm for social inquiry. For he cannot consistently admit that there are any biological laws, unless these laws are in principle deducible from the laws of physics; but the prospects for such a reduction are matters of great controversy among both biologists and philosophers. In the light of such controversy, the physicalist will be at best uneasy about the role here accorded biology.

The empiricist's and the physicalist's doubts about the scientific status of biology are manifested most forcefully in the controversies surrounding the most characteristic of biological claims, the theory of natural selection. Since that theory's promulgation it has been as much a subject of criticism by empiricists and physicalists as it has been a subject of repudiation by emergentists and vitalists. The opposition of the latter has centered on the theory's prospects for eliminating purpose and teleology from our conception of the universe, and substantiating materialist doctrines about the origin of life. Naturally, these features of the theory should endear it to empiricists and physicalists. Yet they have not received it warmly. The principal reason for attacks on the theory from this quarter is the apparent lack of empirical constraints on its explanatory employment and the relative paucity of its predictive content. Indeed, during the century or more since Darwin first broached it, the theory of natural selection has regularly been stigmatized as a grand tautology, untestable, and entirely devoid of any cognitively significant content. Moreover, the theory's employment of notions like adaption, fitness, selection, and their cognates generates for it all the problems associated with teleological language in biological explanations of far smaller-scale events than the theory itself deals with. If these terms are ineliminably teleological in content, then the theory that employs them cannot describe regularities in the behavior of objects which are *nothing but* aggregations of matter obeying the nonteleological laws of physics. As such, Darwinian theory can hardly be accepted either by the empiricist or by the physicalist, for it may turn out to be a body of propositions devoid of empirical content and/or laden with the burden of emergentism. If the theory of evolution is the centerpiece of biological thought, then it may be said that offering biology as the exemplar empiricism invites the social sciences to emulate is more than just surprising, it is patently unwarranted.

Difficulties in the exposition and employment of the theory of evolution are of course recognized by biologists and sociobiologists. In a section of *Sociobiology* entitled "Reasoning in Sociobiology," Wilson notes that it is always easy to provide a retrospective story for any conceivable property of an organism or group of organisms, showing that its manifestation is adaptive or has in the past been adaptive with respect to environmental factors or needs which are difficult to establish or falsify independently. He concludes with the demand that the employment of evolutionary explanations maximize specificity in their citation of these environmental factors and needs, in order to enhance the explanation's testability. He writes, "Paradoxically, the greatest snare in sociobiological reasoning is the ease with which it is conducted. Whereas the physical sciences deal with precise results that are usually difficult to explain, sociobiology has imprecise results that can be too easily explained by many different schemes. . . . What really matters with respect to . . . scientific . . . content is that the statement [be] formulated in a way deliberately to make it falsifiable. . . . A theory that cannot be mortally threatened has little value in science."[9]

But such cautionary advice will not obviate the doubts regularly cast on the theory of natural selection, and the ease with which its employment can violate even self-imposed strictures on testability and empirical content is demonstrated in Wilson's own practice. Thus Wilson writes, "The heart of the . . . hypothesis ["that human behavior is . . . organized by some genes that are shared with closely related species and others that are unique to the human species"] is the proposition, derived in a straight line from neo-Darwinian evolutionary theory, that the traits of human nature were adaptive during the time that the human species evolved and that genes consequently spread through the population that predisposed their carriers to develop such traits. Adaptiveness *means* simply that if an individual displayed the traits he stood a better chance of having his genes represented in the next generation than if he did not display the traits."[10] Unless Wilson's claim about the *meaning* of adaptiveness reflects a careless assimilation of the meaning of a term and the causal consequences of the instantiation of the property it names, his own characterization of adaption makes the theory of natural selection to which he appeals, as essential to his enterprise, quite vacuous. For on this characterization increased rates of survival in successive generations are explained in terms of adaption, that is, the very same increases it allegedly explains. Of course, in his actual practice, and elsewhere in his writings, Wilson makes clear that adaption, fitness, and selection are properties which bear a causal, not a semantic, relation to rates of reproduction. But neither he nor other sociobiologists and biologists have offered an account of these notions which makes their relation to their causal consequences sufficiently clear and precise as to obviate the repeated charge that the theory that employs these terms is a grand tautology, and to preclude the repeated *slip* by its users that transforms it into a body of statements guilty of this charge.

Without an account of the theory of natural selection which explains both why this theory is not guilty as so frequently charged and also why it is often mistaken even among biologists for an empirically empty theory,[11] the empiricist will have difficulty convincing himself, not to mention skeptics, that evolutionary theory is the cornerstone even of biology, let alone the best theory the social sciences can hope for. He must explain why the theory is not guilty of a lack of content for obvious reasons connected with his entire philosophy of science. By the same token, his explanation must show why biologists have misunderstood their own theory, or he will have difficulty convincing them and others that it is *their* theory which he has shown to be innocent of the alleged defect, and not a reconstruction distinct and independent of the theory they actually employ. Not only must these obligations be faced by the empiricist, but the physicalist must also explain how a theory like the theory of natural selection, which makes claims about populations, species, and lineages of individual organisms, can be reduced to one about the anatomy, physiology, and neurophysiology of individual organisms, without passing through an intermediate theory about the individual behavior of these organisms. Physicalism requires that we show at least that there is no obstacle to a reduction of the claims of evolutionary theory ultimately to claims expressible in the theories of chemistry and physics. Our claim that a science employing this theory to explain human events preempts all the conventional social sciences requires that this reduction involve no passage through intermediate laws of the sort the traditional social sciences can claim to provide, and whose existence we have denied.

We may perform all four of these tasks by focusing on the central theoretical notion of Darwinian evolution, the concept of *fitness*. Doing so will involve another excursion into the details of biological theory on which our attention was fixed in Chapter 6. There it was claimed that some of our attention to the details would be repaid subsequently, and this is the stage of our argument at which the promissory notes can be paid. In Chapter 6 the character of the theory of evolution was cited to explain why the appeal to reasons and actions will never eventuate in nomological generalizations. Here the same theory will help us see why the lowest-level practically useful generalizations governing human behavior in which we can be confident are at the level of population biology.

According to the theory of natural selection, differences in the fitness of individuals are a, or the, principal determinant of their differential reproduction, and differences between the fitness of members of different hereditary lineages determine their different prospects of survival. But up and down the phylogenetic scale the only uniform means available for measuring differences in fitness is by measuring differences in past or future rates of reproduction. Misunderstandings of the theory first set in when operationalist imperatives forced biologists to treat this sole uniform measure of differences in fitness as part of the meaning of fitness. Thus, if differences of fitness are defined by reference to differences in *future* rates of reproduction, then fitness cannot be appealed to in explanation of such rates, and the theory in which it figures as the cause of these rates will rightly be judged vacuous. On the other hand, if fitness is defined by appeal to the rates of reproduction of *ancestor* organisms, then although the theory may remain testable, the notion of fitness as an explanatory variable drops out. Thus, consider the following semiformal characterization of fitness:

As a possible operational definition of [the fitness of organism, b], ϕ (b), I might suggest the following. Let $v_1(b,k)$ be the sum over all the k-ancestors of b of the number of reproducing offspring of each. Then let $v_2(b,k)$ be the number of k-ancestors of b. Then $v_3(b,k) = v_1(b,k)$ $/v_2(b,k)$ is an estimate of the average fitness of the k-ancestors of b. Now let the operational definition of ϕ (b) be:

$$\phi(b) = \sum_{k=1}^{n} (0.5)^k v_3(b,k),$$

where n is the number of generations for which data is available. Then about (0.5) of the fitness of b is estimated by the "average fitness" of its parents; about 0.25 of the fitness of b is estimated by the average fitness of its grandparents; etc. . . . $(0.5)^k$ is a factor which adjusts the importance to be attached to more remote generations.[12]

Fitness is characterized by this "operational definition" simply in terms of the reproduction rates of prior members of an organism's line of descent. If fitness is then called upon to explain the present organism's rate of reproduction and the rates of its descendants, then in effect, explanations in terms of fitness are but covert appeals to prior rates of reproduction to explain future rates. The notion of "fitness" thus becomes a *façon de parler* for such prior rates, and has no explanatory standing of its own. But it is clear that in evolutionary theory fitness is not simply a shorthand way of describing past and future rates of reproduction, nor is the theory of evolution committed to the claim that future rates of reproduction are dependent on nothing but past rates of reproduction. Nevertheless, determinations of such prospective and retrospective rates provide the only uniform measure of fitness generally available, and this is what leads to the mistaken view that the theory of natural selection is vacuous.

To see that fitness is a "real" property of organisms, and not merely an eliminable construct, consider two organisms with the genetic identity of twins. Suppose that one has been synthesized from organic material on the basis of a complete genetic map of the other. In this case the synthesized entity will have no ancestors, and the measure of fitness given above for it will be zero or at any rate inapplicable. Since the two organisms are of exactly the same character in physical properties, it must follow that they have the same level of fitness. To deny this is to deny that fitness is a physical property of organisms. Now, destroy one organism (either one) and permit the other to reproduce. Since under these circumstances their rates of reproduction will obviously be different, it follows that differences in fitness cannot be identical to differences in reproduction rates, for *ex hypothesi* the two items were identical in levels of fitness. So, it seems, two organisms can have differing rates of reproduction, both prospectively and retrospectively, and yet have the same level of fitness.

What is more, two organisms can have quite different physical properties and yet share exactly the same level of fitness in respect to a given environment. For example, a bird and a squirrel can occupy roughly the same environment, have exactly the same prospective and retrospective reproduction rates, and yet differ greatly in their anatomical, physiological, behavioral, and environmentally related properties. But how can fitness be a purely physical property of organisms, and not simply a measure of their rates of reproduction, if two such different organisms can have exactly the same level of fitness? By now it should be clear that this sort of case, like the previous one, is possible because fitness is a functionally characterized concept; that is, levels of fitness are the causal consequences and causal antecedents of heterogeneous classes of natural phenomena. (In this respect it is like the concepts of belief, desire, and action that figure in statements like L, of Chapter 5. Unlike those concepts, fitness figures in general statements which we are confident are laws of nature, and so it must characterize a natural kind of property.) Thus, for example, two organisms can have the same level of fitness because one avoids its predators by camouflage and the other by flight, or because one endures severe weather by virtue of its thick coat of fur and the other by migration. In other words, fitness among animals is interconnected to a vast number of different physical properties and environmental conditions. So vast a number of functional concomitants does fitness have, that it would be quite impossible to specify even a small proportion of the nomological connections between a given level of fitness and all of the different properties and relations of organisms that could give rise to it and to which it can give rise. The relations between these items and any given level of fitness are "many-one" and "one-many." These facts explain why fitness functions as it does in respect of prospective and retrospective levels of reproduction, and at the same time why there is no practicable way of measuring fitness differences between differing species (or even different organisms) other than by appeal to reproduction rates. It also explains the difficulty of providing a systematic reduction of the theory of natural selection to any other theory. In fact, to be complete, this reduction would involve the explanation of natural selection by appeal to every theory that governs the particular physical and environmental properties of every species whose evolution natural selection explains. Both of these considerations make manifest the *force* of the conclusion that the theory of natural selection is somehow methodologically suspect: its key term cannot *in fact* and in detail be "cashed in" by appeal to any other theories which it is practically possible to develop, to systematically deploy, or even strategically worthwhile to pursue. On the other hand, these considerations also help show how to state the character of the concept of fitness precisely, in a way that blunts the force of such conclusions and renders perspicuous the logical status of the theory of natural selection.

The concept of fitness is *supervenient* on the manifest properties of organisms, their anatomical, physiological, behavioral, and environmentally relative properties. And this fact alone explains the simultaneous explanatory power and empirical recalcitrance of the concept. Before proceeding to expound the sense in which fitness is supervenient on what I have called the manifest properties and predicates of organisms, I should at least sketch out what sorts of properties these are. By the manifest properties of organisms I mean those properties, dispositions, and abilities which organisms are accorded by theories of anatomy, theories of physiology, and theories about the relations between these properties and properties of the organism's environment. (The properties reflecting this interaction are hereafter called ecologically relative properties.) These properties are the ones to which appeal must be made in any account of an organism's size, strength, speed, longevity, period of sexual maturity, probability of avoiding predators or capturing prey, litter size and frequency, method of feeding young, and so forth. In other words, the types of properties of organisms on which fitness is supervenient are those that figure in nomological generalizations (but not necessary truths) that govern any organism's number of reproductive opportunities and rate of successful reproduction. They are, in short, the causes of its rate of reproduction. Naturally, the instantiation of these properties causally connected to any organism's rate of reproduction will itself be explained by anatomical, physiological, and ecological theories.

The formalization of supervenience employed here is the application of one developed by Jaegwon Kim.[13] Consider the notion of fitness offered by Williams and described above. This functor can take on a continuum of values from 0 to ∞. In effect, these values constitute a denumerable infinity of fitness predicates. Call this set of predicates F. Now consider the set of all the anatomical, physiological, and ecologically relative properties which it is physically possible for an organism to manifest, the natural kinds under which it falls. Call this set P. Let P^* be the set of all properties constructible from P by any combination of conjunctions and disjunctions (and their negations) of finite or infinite length. Among the members of P^* will be properties which are exhaustive of P, in that an organism manifesting one of these exhaustive properties will be fully and completely characterized, both positively and negatively. These exhaustive members of P^* will be mutually exclusive properties of organisms as well. Call this set of exhaustive mutually exclusive predicates constructible from P, P_e. Now the set of properties F, the properties of having various fitness levels, is supervenient on the set of properties P only when it is the case that if two organisms share the same member of P^*—the set of all possible anatomical, physiological, behavioral, and ecologically relative properties—then they share the same property in F, i.e., the same level of fitness. An organism's level of fitness, ϕ, is thus not dependent on or identical with its past or future rates of reproduction, but is a matter of the organism's having a particular set of properties in P. If an organism o has fitness level ϕ_i, then there is a set of properties in P such that o has this set of properties, and anything else with the same set of properties has fitness level ϕ_i. Although fitness is thus determined by features unmentioned in the theory of natural selection, it cannot yet be said to be reduced or reducible to these features. The members of P realized by o are only individually sufficient conditions for particular levels of fitness; they may not be necessary and sufficient; it is biconditionals, stating necessary and sufficient conditions, that are required for reduction. Such biconditionals, however, will be at least in principal constructible from members of P_e, the set of exhaustive and exclusive properties constructed from P, if the members of P are finite in number. To see this, suppose that organism o has fitness level ϕ_i. Then, since fitness levels supervene on properties in P, there is some member of P_e which o manifests, say, P_{eo}, and if any other organism instantiates P_{eo}, then it too has fitness level ϕ_i. But this is tantamount to concluding that each member of P_e, the exhaustive and exclusive properties constructible from P, is sufficient for some level of fitness. If the set of members of P_e is finite, then a finitely long enumeration of the members of P_e, each of which is sufficient for some particular level of fitness, will provide both necessary and sufficient conditions for each level of fitness. Of course, if the number of members of P_e is very large, then such biconditionals will be practically impossible to construct. But the important point is that any particular level of fitness is a function solely of the manifest properties of organisms, and the function in question is that of supervenience. This explains how fitness can be nothing more than having a certain combination of anatomical, physiological, and ecologically relative properties, even though no set of such properties may be stateably necessary and sufficient for a given level of fitness; even though differing organisms in different habitats, with differing prospective and retrospective reproduction rates may (conceivably) have identical levels of fitness; and even though we may be unable to cash any particular level of fitness in for a specific set of such manifest properties of organisms. All these features of the notion of fitness, which its supervenience on manifest properties of organisms brings out, are the very features of a theoretical term in science that has great explanatory potential, just because it is not exhausted by, and cannot be translated into, observable properties.

Consider how fitness, thus understood, figures in the process of natural selection, as a causal variable nomologically connected with, but conceptually independent of, rates of reproduction. The members of line of descent are characterized by a certain set of hereditary, physical, and ecologically relative properties. Together with the environment in which the members of this descent line live, these properties determine a level of fitness for each member measured on a scale between zero and infinity. If the environment changes, physical properties remaining constant, then the fitness number will also change; if mutation changes physical properties, environment remaining unchanged, fitness number will also change. In part, the fitness number is a measure of the "fit" between organism and environment. This fit at any given generation causally determines the number of reproductive opportunities at that generation, and since the fit is partly hereditary, it is also nomologically connected to the number of such opportunities at previous and future generations. This is why rates of reproduction provide an estimate of the fitness number of a line of descent, even though the direction of causation is from fitness to rates of reproduction, and not vice versa. But since the environmental conditions and the physical properties of organisms are so complex, so varied, and so difficult to separate out and quantify as determinants of fitness, fitness can at best be shown to be supervenient on these properties, and we must employ rates of reproduction in order to measure fitness.

It is only by appeal to the supervenience of fitness on physical, behavioral, and ecologically relative properties that we can account for the employment of fitness to explain differences in reproduction rates, and can account for the intra- and inter-species comparisons of fitness that biologists appeal to as the mechanism of successful competition, predator-prey relations, biogeography, niche theory, and other aspects of selection and evolution.[14] A tighter connection to rates of reproduction deprives fitness of its explanatory force. On the other hand a tighter connection than that of supervenience between fitness and the manifest traits of organisms would deprive the notion of its systematic employment in the explanation of how differing organisms in similar or different environments can both survive and compete successfully against other organisms, and how differing organisms can supplant one another in the same environment. Only on the assumption of supervenience can different combinations of manifest and ecologically relative properties constitute the very same level of fitness, and the same manifest or ecologically relative properties constitute differing levels of fitness (*modulo* differences in ecologically relative or manifest properties). Yet it remains entirely consistent with this flexibility in the relation between manifest properties of the organism and its level of fitness, that any organism's *particular* level of fitness *at a given time* consists in, and is identical to, nothing more than the organism's physiological, anatomical, and behavioral properties and the environment in which it finds itself.

Thus, although "fitness" is a theoretical term at the level of the theory of natural selection employed to explain differences in rates of reproduction, levels of fitness of particular organisms are in turn explained by appeal to theories about their physical properties, and the relation of these properties to the organism's environment. This enables us also to see more exactly the relation between the theory of evolution and those other biological and physical theories which, together with the genetic account of heredity, *explain* the theory of evolution. The theory of natural selection rests on these theories, in *something like* the conventional reductionistic picture. Since fitness consists in properties of organisms whose instantiation is deducible from the scientific theories at hand (or in principle available), and since the hereditary transmission of (some of) these properties and the occurrence of variation among them is deducible from principles of population genetics (and its successors), the leading principles of the theory of natural selection should follow from these theories alone. We can sketch out this reduction in connection with the axiomatization of the theory of natural selection provided in Chapter 6. I shall restate each axiom and indicate how the reduction of the axiom to laws in other theories can proceed.

Axiom 1. There is an upper limit to the number of organisms in any generation of species.

This axiom finds its ultimate reduction in considerations from thermodynamics, and needs no special biological underpinning.

Axiom 2. Each organism has a certain quantity of fitness with respect to its particular environment.

Since each organism has a certain proportion of those properties, dispositions, and abilities which are the causal determinants of its number of offspring, and since fitness is supervenient on these properties, this axiom can in principle be shown to follow from the theories which account for why particular organisms have certain proportions of these properties. But because fitness is supervenient on these properties, different organisms can have the same level of fitness in a given environment, and the same organisms can have different levels of fitness in different environments. Since fitness is determined by those properties causally responsible for reproductive opportunities and successes, and since genetic theory (which is separate and distinct from evolutionary theory)[15] assures us that some of these properties are hereditary, it should follow that:

Axiom 3. If D is a physically or behaviorally homogeneous subclass of a species and is superior in fitness to the rest of the species for sufficiently many generations, then the proportion of D in the species will increase.

The fourth axiom asserts that the antecedent of axiom 3 is fulfilled. It is the assertion that fitness is sufficiently hereditary to make for this secular increase in the proportion of D within the entire population.

Axiom 4. In every generation of a species not on the verge of extinction there is a subclass, the members of D, such that D is superior in fitness to the rest of the species for long enough to ensure that D will increase relative to the species and will retain sufficient superiority to continue to increase, unless it comes to constitute all the living members of the whole species at some time.

This axiom should follow from the existence claims of other theories to the effect that among given species there are classes that differentially instantiate the properties causally connected to reproduction, and that some of these properties are heritable.

The question of whether such an account captures all the leading ideas of the theory of natural selection (as its originator, Williams, claims) is a difficult one to answer definitively. Similarly, filling out the reductionistic program here envisioned may be difficult in the extreme. And because the failure of any particular actual attempt to fill it out in no way undercuts our confidence in the theory of natural selection, it is easy, if mistaken, to infer that the theory is a vacuous one. But once it is seen that the program is at least in principle susceptible of completion, the exercise need not be carried out. For merely seeing that the theory of natural selection can be shown to rest on such plainly empirical foundations is all that is required to understand both its power and its conceptual status as an extremely broad-gauged but nevertheless entirely contingent general theory.

The supervenience of fitness on manifest properties of organisms enables us to say that (1) the state of a particular organism's manifesting a given level of fitness at a given place and time is identical with, and is nothing but, the state of its exemplifying a certain set of manifest properties, even though (2) in general the property of having that level of fitness is not identical with or extensionally equivalent to the complex property of having those particular manifest properties. Thus, a particular exemplification of fitness is subject to just the (qualitatively and numerically) same causal and mereological forces as the instantiation of a particular set of manifest properties is, even though two distinct instantiations of the same level of fitness are each likely to be identical with instantiations of different sets of manifest properties. Thus, even though the terms of the theory of natural selection may not be reduced to those of more fundamental theories, these terms are employed in the theory to describe and explain the very same states, their causes and effects, as the terms of more fundamental theories; and if the more fundamental theories provide more fundamental (if less manageable) explanations of the same states, then in and of itself the failure to provide reduction functions for the terms of the theory of evolution cannot cast doubt either on the scientific legitimacy of this theory or on its consistency with the thesis of the unity of science.

Treating fitness as a property of organisms supervenient on their physical and environmental properties simultaneously underwrites the conceptual independence of fitness from rates of reproduction while recognizing its operational dependence on such rates. Keeping this distinction between the meaning of a term and the operations for determining its application clearly in mind enables us to see that the theory in which the notion of fitness plays so great a role is not a vacuous one, in spite of the fact that the kinds of variations which it explains represent the only kinds of phenomena available for specifying its explanatory variables. Insofar as the empiricist can make this distinction between meanings and operations, he has the resources to explain why the theory of natural selection is a legitimate empirical one, and why biologists mistakenly treat it sometimes as an empirically contentless one. Moreover, recognizing the supervenience of fitness on the properties that figure in the laws of physics and chemistry enables the physicalist to at least sketch out the reduction of the theory of natural selection to these more fundamental theories, and thereby preserve his version of the doctrine of the unity of science. And since the sketch appeals to no laws intermediate in generality between those of physics, chemistry, and a theory of genetics reducible to the former, the physicalist's reduction sketch substantiates the independence of the theory of natural selection from laws of animal behavior akin to or identical with those figuring in conventional behavioral and social science. The reduction preserves, therefore, the preemption of social science that sociobiology requires.

In many respects this analysis is just what the empiricist and the physicalist need to assure themselves of the respectability of the theory of natural selection. But in other respects, and, in particular, for our purposes, it might appear that our argument has proved too much. That is, a defender of the existence of general laws, and of autonomous sciences and theories characteristic of animal psychology or the human sciences, might complain our argument either reflects a double standard, clothed in the technicalities of "supervenience," or provides enough grounds to reintroduce those kind terms like 'reinforcement,' 'operant,' or indeed, 'belief,' 'desire,' 'action,' which sociobiology proposes to supersede. For what does the analysis of fitness in terms of supervenience amount to if not the admission that no finitely stateable specification is available for this term which is also independent of the phenomena it is employed to explain? If this is the case, it may be argued, then fitness is no more and no less legitimate a notion than these others, and the theories in which they figure are at least as respectable as the Darwinian one which seeks to preempt them. The antecedent of this conditional can hardly be gainsaid. For it does appear that at best the differences between complexities of independent specification among these terms are ones of degree. But the consequent of the conditional does not follow. It should be clear that the argument mounted in this book does not conclude that terms like 'belief' or 'desire' or 'action' are logically incoherent or unintelligible; the argument constitutes an explanation of why social sciences have uncovered no laws employing these terms. Part of the explanation involves revealing the lack of independent specification for them and accounting for this failure to be independently specified by noting the spatiotemporally restricted meanings of these terms. The philosophical advantage of this explanation is that it provides an analysis of the failure of empirical methods to produce results in social science which is (1) compatible with the correctness of these methods and (2) far more plausible than other compatible explanations.

Lack of stateable independent specification for the terms of a general statement are neither necessary nor sufficient conditions for that statement's failure to be a law, but they do cast suspicion on its nomological status. That is why empiricists have been dubious of the theory of natural selection. Under some conditions, the lack of such specification is sufficient to deprive a statement of lawlike standing. But such circumstances do not obtain in connection with the theory of natural selection. To see this, consider differences between this theory and our exceptionless general statement about reasons and actions, L. Fitness, the supervenient and therefore allegedly unspecified explanatory notion of the theory, figures not in one but in three distinct general statements of that theory, each of which provides a means logically independent of the others for determining fitness, and each of which will be a law if the theory is embraced. Moreover, these statements have logical consequences, such as the theorem of Chapter 6 to the effect that the population levels of competing species must approach an equilibrium value, in which the notion of fitness does not figure, and whose application to particular cases does not require the description of these cases in terms that appeal to the notion of fitness. Finally, while there are cases in which the explanatory application of the theory of natural selection is, like the application of L, uncontrolled by independent means of establishing initial conditions, there are other cases in which the explanatory and predictive application of the theory is independently controlled, and the theory is thereby confirmed. For instance, given species like fruit flies or bacteria that breed rapidly in comparison to mammalian rates, we can arrange laboratory circumstances in which fitness is clearly specified without appeal to reproduction rates, and in which these rates are then explained in quantitative detail by the theory, and predicted by it as well. Finally, the fact that the theory requires a mechanism of heredity, and the fact that such a mechanism has been discovered and made the subject of well-confirmed laws plainly independent of the theory of evolution itself, is further reason to distinguish its nomological status from that of L, in spite of the apparent similarity between its theoretical concepts and L's. In these respects, the operant theorist's law of effect stands somewhere in between natural selection and L in its claims on our confidence. It can be applied to cases in which the initial conditions are independently though only species-specifically specifiable, the events it explains can be described independently of its own descriptive terms, and its explanatory terms seem definable if not specifiable independently of the members of particular species whose behavior it explains. On the other hand, the claim that its explanatory property, reinforcement, is supervenient on purely physical or chemical ones is clearly more speculative than the analogous claim in evolutionary theory, and the theoretical ramifications of the law of effect are clearly not as considerable as those of Darwinian theory. In the end, the fact that our account of fitness and its role accords some analogical creditability to reinforcement theory tends to strengthen our argument both because of the similarity between these two theories of the natural selection of species and of behavior, respectively,[16] and because our earliest arguments independently accorded the law of effect a chance to demonstrate its nomological standing.

The empiricist's search for the lowest-level generalizations and the narrowest natural kinds under which we can be confident that human behavior falls leads him to a sympathetic consideration of the claims of sociobiology. Both the sociobiologist and the physicalist recognize that neurophysiology, with its resources in chemistry and physics, can provide explanations and predictions of particular instances of behavior that are in principle as complete and precise as the laws of chemistry and physics provide in their own domains. Yet both also recognize that at the level of complexity characteristic of primates, the specifications of initial or boundary conditions for the neurophysiological explanations and predictions of this behavior require more information and industry than we can foreseeably expect to amass and deploy within the time frame available for prediction or the cost constraints of satisfying our explanatory curiosity. Neurophysiology will not supplant the social sciences because it cannot satisfy the need for which they hope to cater within the allowable costs that the users of social theory set. The sociobiologist's claim on our attention, on the other hand, rests in part on denying the possibility that these subjects will ever succeed in filling this need, and therefore claims the right to preempt them. But the plausibility of this claim to preempt seems to involve the outright denial that our predictive and explanatory demands will ever be satisfied within the limits set for them. To the degree that these limits on information available and deployable are reflected in our attachment to the explanatory language of beliefs and desires as the only variables inferable from behavior and our attachment to intentionally characterized actions as the events to be explained and predicted, the denial that our demands will ever be met made by sociobiology is equally a consequence of our empiricist attempt to account for the failures of social science traditionally conceived. To the extent that the empiricist can embrace the leading theoretical idea of sociobiology, this convergence lends credence both to its explanation of the failures of social sciences and to sociobiology's claims to preemption. But this congruence still leaves open the important question of what exactly we may hope for from the pursuit of a sociobiological program of research, sustained, as it now appears, by the confidence of empiricism.

8

Using and Abusing Sociobiology

What, in fact, will a theory be able to tell us of interest and of use if that theory reflects the assumption that the closest we can get to an explanation of animal behavior is an account of the distribution and transmission of the heritable traits which it reflects? The answer to this question is clearly contingent on empirical research and will vary enormously from species to species. Thus, there are species of mosquitoes all of whose members' behavior is identical in sequence, duration, and topological configuration, and is determined solely by their common genetic inheritance. For such organisms a sociobiological theory tells us everything there is to know above the level of neurophysiology. There are other, more complex species, like ants and bees, whose aggregate behavior seems equally fixed and whose "social organization" is equally amenable to complete sociobiological explanation. As a matter of logic, there is no limit to the potential informativeness of a theory that accounts for nothing closer to individual activities than the dispositions that enable organisms to participate in these activities. For obviously, if the dispositions are so narrow and so determinative of the occurrent behavior they allow, explaining them explains everything else as well. As a matter of scientific theory, the only limitations to a sociobiological theory's explanatory application are the levels of dispositions and of occurrences that are heritable phenotypes, and the degree to which their contributions to individual and species fitness can be determined. As a matter of the factual beliefs of sociobiologists, the limitations on application of the theory to occurrences and to dispositions of organisms appear to increase as we move up the phylogenetic scale (although there are important peaks and valleys in this curve). And as a matter of practical utility, the explanatory limits on this theory for individual human dispositions and behavior are limited by the great difficulty identifying phenotypes above the level of biochemistry. Thus, the chapter of *Sociobiology* devoted to applying the theory to human behavior, "From Sociobiology to Sociology," is remarkably free from firm claims about either individual behavior or social institutions.[1] Indeed it reads like a sustained argument for the impossibility of explaining much about either individual or group behavior on the basis of evolutionary considerations. Thus, Wilson notes that "the parameters of social organization vary far more among human populations than among primates"; this difference is attributed to greater individual differences in ability, intelligence, and personality among humans and, according to Wilson, suggests that genes which promote flexibility in social behavior are selected for at the individual level. In other words the phenotype that makes for social flexibility has proved fitter than others, and this has restricted the application of the theory below the grossest common levels of social organization. Another explanation for this variation and the consequent explanatory or predictive weakness of the theory lies in the comparative lack of competition humans face from other species. These conclusions should by now be expected, for sociobiology's claim to preemption is based on the hypothesis that this genetically allowed variation in behavior is so great that no stateable laws about individual behavior can be discovered. These admissions also make particularly absurd the condemnation of sociobiological reasoning by attributing to it an absurd thesis of biological determinism. If anything, this theory is wedded at the level of molar human behavior to a thesis of hereditary indeterminism. Bereft of his commitment to a thesis of explanatory biological determinism, Wilson's claims sound more like an argument against sociobiology than one in favor of it, for he writes that "only in man has culture so thoroughly infiltrated virtually every aspect of life that ethnographic detail is genetically underprescribed, and consequently so highly diverse."

At the least sociobiological theory and its basis have this use: they can curb our expectations and explain why in many cases our demands on social and behavioral scientists and "engineers" have not been satisfied. The explanation does not reflect on the intelligence or industry or sincerity of these people, but on the impossibility of our demands. It will be no surprise that we cannot predict the effects of a piece of social welfare legislation, or a media advertisement campaign, or a politician's promises, or a mother's child-rearing practices, up to the level of accuracy we need in order to know in advance whether the good effects of the legislation will outweigh the bad, whether the sales will respond to the costly sales pitch, whether the politician should change his tune, or whether the child is right to blame his mother for his problems. We cannot formulate generalizations that will enable us to make such predictions because our attempts will always be couched in terms that do not pick out natural kinds; we will inevitably lump causally heterogeneous factors together because they meet the same nonqualitative descriptions. And if the heterogeneity is as great as the empiricist and the sociobiologist require to diagnose the failure of social science and preempt it, then no amount of further refinement will reduce the error term of the statistical generalizations, to which the practical, social, and behavioral scientist appeals, to levels required for reasonably complete explanation and reliable prediction.

Thus, when we employ an econometric model in order to predict how a change in the money supply or the interest rate or the budget deficit will affect the employment level or the inflation rate, we do so on the assumption the relation between these variables will be mediated by the effects of the economic changes on the beliefs and consequently on the actions of economic agents. If there are no regularities relating the mediating variables of belief and action, it will be no surprise when the econometric model gives the wrong results. Naturally, it is better to employ the models at hand rather than no models at all in the light of the fact that we must make the most intelligent decisions we can about such matters. Moreover, we can even employ the model while specifically rejecting its disaggregative presuppositions, treating it as a computational device, and not the reflection of the economic theory that generated it; alternatively we may argue that although there are no deterministic generalizations underlying the model, either about reasons and actions or about any other intermediate links in the chain from fiscal or monetary policy to economic statistics, the variations among underlying variables are sufficiently small to average out over all the agents in an economy. But although each of these ploys helps justify our use of the model, none of them gives reason to suppose that we can improve it up to arbitrarily selected levels of accuracy, none of them provides the model with more secure standing than that of an accidental generalization, and none gainsays the sociobiological caution that we should limit our expectation of its accuracy, and not demand of it greater accuracy in prediction than an accidental generalization admits of.

When it comes to models and theories at the other end of the scale of aggregation—the ones that purport to account for the determinants of individual action, that are employed to diagnose and treat individual behavioral disorders, and on the basis of which we make judgements of ability, disability, condemnation and punishment, praise and reward, guilt or innocence—the negative usefulness of the sociobiologist's and empiricist's conclusions are even greater. Theories like psychoanalysis and its welter of variations and alternatives have long been condemned by empiricists as empty and untestable, or completely discredited by their practical failure to effect "cures" or even improvements in psychiatric patients at a rate more frequent than spontaneous remission. If attempts to change behavior, by adding to or changing patients' desires, fall afoul of the lack of nomological connection of behavior with states so characterized, then it is no wonder that such theories and their associated therapeutic regimes must fail. Our theory enables us to explain the hold which such theories have over both laymen and therapists, and the hope, successively frustrated, that the latest of them will really provide understanding and effect improvement. The hold is a reflection of our conviction that most of our singular statements about the determinants of action are true, conjoined to the natural assumption that such singular statements reflect regularities expressed in their own terms. Seeing the consequences and the vanity of this last assumption loosens the hold of each of these theories and dampens the hope that any of them could be correct or could work at a level of success above chance.

Our account makes equally vain the expectations of the anthropologist and the historian that their studies of particular cultures or particular episodes should eventually help provide an understanding of cultures in general or the course of history. The hope that we might be able to plot history's future course by learning its past lessons and the aphorism that failing to do so forces us to repeat them have sometimes been stigmatized as "historicism," and have been traditional targets of empiricist condemnation.[2] As is the case for speculative theories of clinical psychology, such claims and the particular theories of history that march under their banner have been in the past condemned as empty of content and untestable. In both cases, the surrender by philosophy of any criterion of cognitive or empirical significance has given psychoanalytic and historicist doctrines a new lease on philosophical respectability. Empiricism's new alliance with sociobiology provides it with fresh resources to combat these doctrines, for the alliance helps explain the impossibility of laws of the sort the psychoanalyst or the historicist requires, instead of simply condemning their claims to cognitive insignificance. The case is somewhat different for anthropology. Anthropologists often justify their study of other cultures simply by reference to its intrinsic interest. With such a rationale, their findings are no more open to criticism than they are to scientific systematization. But if anthropologists hope to found on the basis of their research a general theory of culture, replete with nomological generalizations and the other trappings of science, they cannot expect to succeed. Now there are few anthropologists who actually embrace such a goal. But many purport to reconstruct the relations among cultural institutions of diverse peoples as reflecting a common structure whose elements are held together by isomorphically held shared meanings. This, for instance, seems the only view attributable to Needham.[3] For in eschewing the selectionist interpretation of Lévi-Strauss's doctrines, he leaves himself nothing more than the web of meanings accorded to acts, institutions, and artifacts in order to tie them together. Anthropologists typically generalize this web of meanings into the network holding the constituents of all cultures together, and diagnose the successes and failures of cultural transmission in terms of the preservation of these meanings. This sort of metatheoretical claim justifies both their frequent sojourns among native people, and the relativism with which they temper (and sometimes altogether melt) their judgements about the subjects of these visits: only by going native can meanings be learned, and only within the circle of indigenous meanings can reasoning be assessed. We can now see that the constant disagreement among anthropologists about the meaning of a given kind of institution both within and beyond a specific culture are never likely to be settled. Quine argues, in *Word and Object*, that these meanings will always be underdetermined by the behavioral evidence for them; what is more, we may add, they will not bear even a nomological relation to any kind of behavior natives display. As such, the circle of meanings into which a given culture is molded or into which all cultures are molded by the anthropologist can have no more than an accidental success in guiding our decisions about how to deal with alien cultures. Indeed, the idea of culture turns out to be as systematically sterile a notion as the anthropologist's intentional notions of "meaning," "belief," and "desire." Although anthropologists may please themselves to construct symbol systems and interpretive universes, from our perspective their construction will be arbitrary, and their ascription, without reliable consequences. On the other hand, those anthropologists adopting a functional perspective, searching for universals, or at any rate, institutions that play common roles in differing cultures, will be vindicated by our view. For they at least are searching for phenotypes at the level of interbreeding populations whose distribution and persistence is explicable by the laws of natural selection. This brings us to the positive uses of sociobiology.

Aside from curbing some of our expectations, the theory can also satisfy others. For clearly, insofar as humans are biological systems, the theoretical claims and mathematical models constructed by ecologists, population biologists, and zoologists are in principle as applicable to them as they are to other creatures. Thus, the Lotka-Volterra equations of Chapter 6 are as applicable to two reproductively segregated but interacting human lineages as they are to nonhuman predators and prey or hosts and parasites. The existence of equilibria levels for two such populations is guaranteed by the theory of evolution, and the divergence from such levels over a long run will reflect significant changes in the relations between the two lineages. These considerations immediately raise questions concerning the relations among castes or classes, between master and slave societies—indeed, a whole panoply of questions which have exercised sociologists and political scientists as well as historians, questions which sociobiological theory holds out the hope of providing a nomological framework for, and the hope of adapting the mathematical expression of its theories to the treatment thereof.

Naturally, applying such a model to the systematization of relations among properties like population levels among reproductively isolated groups presupposes that such properties behave like or are related to other properties which behave like genetic phenotypes. But there is no way to assess this presupposition except by employing it and examining its explanatory power. Of course, the direction of reasoning can be reversed. If demographic evidence shows fluctuations in rates of reproduction for social groups that are characteristic of those derivable from a predator-prey model, then the question of whether these groups bear such a relation to one another can be raised in a context where the dispute has some hope of solution. Similarly, among population ecologists, a sequence of models has been offered to account for the relative rates of reproduction of competing species. Such models may ultimately be applied to explain disappearance of human lineages, or again, oscillations among the populations of competing groups. Thus, Gause's principle, that two species with identical requirements cannot coexist in a habitat, not only explains much about the distribution of finches in the Galapagos,[4] but may also explain the occurrence of a variety of human displacements, migrations, extinctions, and conflicts, by deduction from the theory of evolution instead of by speculation on the character of the competing agents. Similarly, theories of the migration, spatial distribution, and territoriality of organisms of all sorts provide ready models for the explanation of the like phenomena among humans. Studies of the factors affecting the numerical diversity of species and lineages in a region, or the general conditions on the stability of the population of a region composed of any number of different species, seem equally applicable to the explanation of the demographic interaction of reproductively separated human groups.

One important aspect of the application of biological theory to the explanation of genetically assimilable features of human behavior is its demands on mathematical relationships between quantitatively expressed hypotheses. I emphasize this because the level of formalism, always treated as a hallmark of developed science, is extremely high in biological theory, and the character of the demands on mathematics is identical to the demands made upon it in the most quantitative of the social sciences, economics. Economists and social scientists generally are ignorant of these two facts. This ignorance is important to correct, not because employment of relatively inaccessible mathematical niceties is required by any respectable science, but in order to show that its employment by biology reflects that subject's degree of theoretical advance, and that much of the formal mathematical methods that now play a role in economics especially, are already relevant to and will become increasingly important in a biological theory of human behavior. To the degree that economists demand that powerful theory be couched in the language of differential calculus, or linear algebra and topology, to the degree that they focus on the mathematical conditions for general equilibrium of systems as required for the descriptions of such systems, to these degrees they will have to admit that biological theory is at least formally as powerful as their own. And when we add its undoubtedly greater degree of confirmation to the fact that economic agents are governed by its laws as well, the appeal of this theory as a subject for application to the phenomena hitherto treated in social science is obvious.

The employment of similar formalisms by economic theory and biological theory should be no surprise. As noted in Chapter 4, both these theories are extremal ones, both direct us to search for equilibria, and this is what enables both to be expressed in terms of differential calculus. It should also be no surprise that just as economic theory has found it more convenient to state its claims in the language of topology and to prove powerful results, otherwise unattainable, by employment of fixed-point theorems, so too ecological theory has come to be expressed in the same way. Indeed, it has borrowed from the mathematical successes of economics to answer its own formal questions. Thus, in at least one standard text on population biology,[5] the criteria for qualitative instability of the population of a group of arbitrarily many interacting species are listed without proof (although the biological significance of each criterion is explained), and the reader is directed to a paper in an economics journal by two mathematical economists[6] for a proof of their sufficiency for instability. (Parenthetically, we may note that if the purely abstract studies carried on by mathematical economists—which plainly are not motivated by prospective application in economic policy, and are more properly classified as studies in mathematics—ultimately find their employment in biological theories, then they will turn out to have been as valuable and important as the apparently useless nineteenth-century studies of non-Euclidean geometry and Tensor calculus turned out to be for twentieth-century physics.)

Consider an example of the power of biological theory to explain facts which anthropologists returning from the site of their investigations usually explain in terms of the myths and meanings with which their subjects' languages are ladened. Suppose we wish to know why in a lush semitropical environment there are two tribes which both hunt only one among the vast numbers of available animals in their environment; and, what is more, in spite of the general abundance in their region depredate one another regularly and almost to extinction. To explain these facts we may have recourse first to a consequence of evolutionary theory that can explain why species in resource-unlimited environments become specialists, depending on only one of the available resources. Then we may appeal to another consequence explaining why two such species compete in ways that do not produce an equilibrium level of population. The following mathematical account of specialization is adapted from one formulated by J. Maynard Smith,[7] and shows that on evolutionary assumptions a species exposed to abundant though difficult-to-catch prey of different kinds will be forced by selective pressures into specialization in its consumption of prey. If n_j is the number of attacks a predator makes on resource j, per predator per unit time, then

$$n_j = KR_j p_j (1 - \Sigma n_j D_j),$$

where K is a constant, R_j is the density of the j-th resource (its level of population), p_j is the probability of attack on the j-th resource when encountered, and D_j is the time consumed between the beginning of the attack and the moment the predator begins to attack again. It is assumed that the number of encounters with prey is proportional to j, its density. The rate of increase in the predator's population is then given by

$$dx/dt = x(\Sigma n_j W_j - T),$$

where W_j is the expected gain, measured in units of x, the density of the predator species, from an attack on resource j (allowing for the possibility of failure to trap j), and T is a constant measuring the food required to maintain the individual. Selection operates on p_j, W_j, and D_j, for these variables reflect the organism's ability to satisfy his food needs. Manipulating the equations reveals that dx/dt, the rate of change in population, will be at a positive maximum, the one required by the theory of natural selection, when the following fraction is maximized:

$$\frac{\Sigma p_j R_j W_j}{\Sigma p_j R_j D_j};$$

and this fraction's value is maximized if the value of $p = 1$ for that resource of which the value of W_j/D_j (the probable gain from an attack per unit of time spent between onset of attacks) is greatest, and the value of $p = 0$ for all other resources. On the interpretation of p_j, this means that a maximally adapted predator satisfying these equations will always attack one species when encountered, even though there are others available (the value of p for these is zero), because it provides the greatest chance of supplying his needs per unit of effort or time available to satisfy them. Naturally, the explanation does not commit the hunter to actually perform such calculations, even though it appeals to the calculations, results as the determinant of behavior. Here is another important contrast between sociobiology's and economic theory's employment of maximizing hypotheses. The explanation here is clearly teleological, and in this respect like an explanation the economist would offer for the action of an agent; moreover, both explanations involve the hypothesis that the behavior occurs because it maximizes some future, concurrent, or standing quantity. But the Darwinian explanation's strength lies in its power to eliminate this teleology by appeal to an underlying non-teleological mechanism of variation and selection. The economic explanation's teleology is a function of its appeal to desires and beliefs, and is not eliminable. Moreover, the acceptance of the economist's explanation, in the face of the fact that agents do not report undertaking the calculations it implicitly attributes to them, turns on the plausibility of some special pleading or other which insulates the explanation's assumptions from test. The evolutionary explanation accords no intentional states to its human subjects, any more than it accords them to its nonhuman ones. In short, the sociobiologist's explanation seems to have all the formal and methodological advantages of the economist's with none of its shortcomings.

The surprising result that abundance and selection force adaptive species into the role of specialists, and not generalists, in the exploitation of resources, now enables us to explain why two tribes should engage in recurrent warfare even in circumstances that appear to provide enough resources for both. The Lotka-Volterra equations permit us to model relations not only between predator and prey, but also between competing species. Consider the following version of these equations:

$$dx/dt = x(a - bx - cy)$$
$$dy/dt = y(e - fx - gy)$$

where a and e are the intrinsic or independent rates of increase of the two species, whose densities are given by x and y, b and f are the inverse of the carrying capacity of the environment for each species alone, and $-cy$, $-gy$ describe the inhibiting effects of each species on the other. Considerations of a kind well known in the theory of general economic equilibrium reveal that if the intrinsic growth rates for the two groups are equal, then their populations will reach a level of stability if $b > f$ and $g > c$—that is, the increase in numbers of either group inhibits its own growth more than it inhibits its competitor's. In the present case, $b \leqslant f$ or $g \leqslant c$ if both compete for exactly the same resource, and the conditions for population equilibrium cannot be met. Accordingly, one group must eventually either drive the other out of existence, or at least reduce the other to a level at which it can no longer compete at effective levels. Here we have an example of the operation of Gause's principle, cited above, to the effect that two species with the same needs cannot coexist in the same environment. Naturally, if the fall in the population of the declining group has negative effects on the values for p_j, R_j, W_j, or D_j for the increasing group, then its maximal rate of increase may decline and may even become negative, allowing an increase in the levels of population of the smaller group and enabling the cycle to repeat itself.

The cultural anthropologist may wish to explain this cycle in terms of prey sacred to tribes and aliens whose own predation defames the prey, in order to account for the anomaly of specialization and aggression in the midst of apparent abundance, but his explanation will be nomologically isolated from the explanation of other anthropological findings, and not expressed in terms that make its general explanatory principles open to empirical assessment. Anthropologists, as we shall see, are often indifferent to these considerations, but anyone already committed to empiricist strictures on explanation will prefer one of the form just sketched. Here, at any rate, is a positive example of the sort of phenomena that a sociobiological approach to human behavior may account for, by employing principles originally formulated without thought to human applications, and warranted by a body of nomological generalizations that we have independent reason to embrace.

Examples—whether hypothetical, like this one, or actual—are, of course, not as decisive as they might be, for, as sociobiologists and their critics seem to agree, the real test of this theory is whether it can explain social institutions and individual traits that are alleged to be unique to human species and, what is more, apparently incompatible with the mechanism of selection that sociobiological theory relies on. Paramount among these traits whose explanation is agreed to be crucial to the plausibility of the theory are the almost universal dispositions that go under the name *altruism*. We noted in Chapter 4 that Becker refers to the phenomenon of altruism as "the central problem" for sociobiology and offers as an argument in favor of his theory that it solves the problem on a limited scale in a way consistent with the hypothesis of individual fitness maximization.[8] The centrality of the problem of altruism is not accorded by Becker, but only reported by him. Wilson too, in the earliest pages of *Sociobiology*, argues that the central theoretical problem of his science is how altruism, which by definition reduces personal fitness, can possibly evolve by natural selection.[9] Of course, one of the reasons that sociobiologists accept the problem of altruism as a crucial test case of their theory is their belief that they can adequately explain the phenomena that go under that name; and the methods they use to do so cast light on much else that is apparently unique to human behavior. Thus, humans show the disposition to obey rules, even in cases where the rules cannot be effectively enforced and where obeying these rules involves costs to the fitness of the individual, or to some plausible surrogate, like food, comfort, pain avoidance, and the like. The social roles and institutions constituted or subserved by such unselfish behavior seem anomalous on the assumption of individual fitness maximization, and their origin in the behavior of individuals or small groups with no expectation or power of reciprocation seems quite incompatible with the operation of selective forces. Altruism, with its supererogatory connotations, is simply a reflection of the ethical systems that are ubiquitous across human history and culture. By and large, these systems, imposed by humans on themselves, sometimes enforced by sanction, sometimes not, constrain individuals to undertake actions at variance with the dictates of self-interest. As such, they represent a widespread property of the human species not immediately open to Darwinian explanation. Or so it appears. Sociobiologists, as I noted, accept the challenge posed by this apparent counterexample. Indeed, their claim to the attention of social scientists can be dated from the onset of their conviction that they had solved the theoretical problem of rendering the existence of altruistic behavior at least formally consistent with the exigencies of evolutionary theory.

It is important to recognize that merely showing this theoretical possibility is as much as we can demand of sociobiological theory. We should not expect at this point the deduction of the occurrence of particular types of altruistic behavior from the principles of natural selection; still less can we complain of sociobiology's inability to explain the details of and differences among particular manifestations of altruism in social institutions, roles, and rules of behavior. Wilson recognizes that such expectations reflect "a common objection raised by many social scientists," and he grants "at once that the form and intensity of altruistic acts are to a large extent culturally determined. Human social evolution is obviously more cultural than genetic. The point is that the underlying emotion, powerfully manifested in virtually all human societies, is what is considered to evolve through the genes. The sociobiological hypothesis does not therefore account for differences among societies, but it can explain why human beings differ from other mammals and why, in one narrow aspect, they more closely resemble social insects."[10] Of course, given the thesis of explanatory biological determinism attributed to sociobiologists in Chapter 7, the claim that human evolution is more cultural than genetic is the tacit assertion that no systematic explanations are possible for the vagaries in the form and intensity of culturally determined altruistic acts. But to say that sociobiological theory cannot account for these differences among societies is to say that nothing can. For consider, without constraints on explanation imposed by the thesis of explanatory biological determinism, what is this claim but the admission that there is a large, indeed crucial, area of social behavior that sociobiological theory cannot explain, but which a study of culture might? Under these circumstances, showing the consistency of altruism and evolution is little more interesting than showing the consistency of altruism and gravitation. If we are firmly committed to the existence of both, we are *ipso facto* committed to their consistency already and will learn little new from a proof of it. Moreover, since altruistic and other forms of morally assessable human behavior are characterized in terms of the non-natural-kind language of reasons and actions, it follows that they cannot figure in general laws and that the phenotypes whose distribution and transmission evolutionary theory explains cannot include such behavior or the dispositions it reflects, *under their normal descriptions*. Accordingly, we cannot expect sociobiological theory to do more than account for the possibility of behavior we can only characterize in such restricted, nonqualitative terms.

This is reflected in the very definition under which the behavior to be explained is described. The sociobiologist defines an organism's behavior as altruistic if the behavior increases the fitness of other organisms at the expense of the organism's own fitness. Now, although much human behavior can be seen to be comported by this definition, it provides conditions neither necessary nor sufficient for altruism as ordinarily understood. For one thing, as ordinarily understood the notion involves the suggestion that the altruistic agent has certain intentions, and that his behavior is not just accidentally beneficial to others and costly to himself. The sociobiologist's definition does not reflect his agnosticism on the question of whether altruism is intentional or not (in this respect, his definition ignores intention for reasons different from those involved in Durkheim's definition of suicide[11]); it reflects his presupposition that altruistic behavior cannot be accounted for in terms of its determination by altruistic intentions, because no behavior can be nomologically accounted for in terms of intentions. In short, sociobiology's incapacity to explain altruism in the detail a social scientist would like is a function, not of its weakness, but of the impossibility of any theory's satisfying the social scientist's demand. The most we can hope for is to see how behavior we ordinarily describe as altruism is compatible with selection of the fittest, by seeing that the theory, together with certain initial conditions, does entail the existence of altruism under its own special definition.

One unit of selection, we have seen, is the individual organism, while the unit of evolution is the line of descent, the species. But the species too is a spatio temporal individual, and insofar as it is capable of dominance or extinction it too is a unit of selection. Not only will the fittest among organisms survive and consequently reproduce in greater numbers, but the fittest among species will do so as well. Thus, if individuals within a species are altruistic in the terms of the sociobiology definition, but this altruism somehow increases the fitness of the entire species, then in the long run hereditary tendencies to altruism may come to predominate in the fittest of species. The mechanism that leads to the spread and ultimately the fixity at some optimal level of such a phenotype within the species is given in genetic theory. Suppose that variation results in the appearance of an individual with the required phenotype, which that individual passes on to at least some descendants before committing a fatal altruistic act. If this act increases the fitness of another individual whose own survival enhances the long-term fitness of the species, including the descendants of the altruistic individual, themselves genetically predisposed to commit altruism in the future, then in the long run the predisposition to altruism may be retained in the species. Wilson expounds this sort of reasoning in a quantitative form as follows.[12] Having characterized altruism as above, he defines "inclusive fitness" as the sum of the fitness of the individual and of all the effects of changes in his fitness on the fitness levels of his genetic relatives. Consider r, a coefficient of genetic relationship, reflecting the proportion of genes any two organisms have in common. Thus for two brothers, x and y, $r(x,y) = \frac{1}{2}$; for uncle and nephew $r(x,y) = \frac{1}{4}$; for cousins $r(x,y) = \frac{1}{8}$. If k is a ratio of the gain in inclusive fitness produced by an altruistic act to the loss in individual fitness produced by that act, then genetically based acts of altruism will evolve if the average inclusive fitness of the individual member of an altruistic species is greater than the average inclusive fitness of nonaltruistic species: that is, if $k>1/r$, the average of the coefficients of relationship within the species. The evolution of altruism is thus described as a consequence, not of individual selection, but of "kin selection," defined as "the increase of certain genes over others in a population as a result of one or more individuals favoring the survival and reproduction of relatives who are therefore likely to possess the same genes by common descent."[13] Kin selection will not only explain the occurrence of altruistic acts among members of the same and closely related lineages; it will also explain the increasing infrequency of altruism among increasingly genetically separated members of the same species.

Kin selection, and the possibility of altruism that it permits, is but a small portion of the theoretical edifice of sociobiology. That theory also includes accounts of social diffusion, migration, territoriality, long-term demographic phenomena (such as the age structures of species and subspecies, their spatial distribution, rates of diversity, competition and specialization), dominance, roles, castes, sexual relations, and parental care. Yet the issue of kin selection and its upshot for the preemption of social science by sociobiology has become a central focus of debate, all out of proportion to its theoretical importance. This misplacing of the focus reflects the willingness of sociobiologists to allow their opponents to determine the location of the dispute between them. Moreover, it reflects misunderstandings of sociobiology's claims by its opponents, and of the theory's presuppositions by its proponents. The most sustained example of these misunderstandings is reflected in the work of an anthropologist, Marshall Sahlins, whose attack on the very possibility of sociobiology, *The Use and Abuse of Biology*, reflects many of these misconceptions.[14] Examining his arguments reveals that clarifying these misunderstandings of sociobiologists about their own theory, and of its opponents about what they oppose, does much to defuse the debate over kin selection as a test case for sociobiology.

Sahlins' book begins with a "critique of vulgar sociobiology." The critique seems to turn on attributing to sociobiology the strong thesis of biological determinism described in the last chapter: sociobiology is based on "the principle of the self-maximization of the individual genotype, taken as the fundamental logic of natural selection. . . . The chain of biological causation is . . . lengthened: from genes through phenotypical dispositions to characteristic social interactions. . . . The idea of a necessary correspondence between the . . . two, human emotions or needs and human social relations, remains indispensable to the [sociobiological] analysis" (p. 4). There is, on the sociobiological view, "a one-to-one parallel between the character of human biological propensities and properties of the human social system." The "Durkheimian notion of the independent existence of social facts is a lapse into mysticism.

Social organization [is] . . . the behavioral outcome of the interaction of organisms having fixed biological inclinations" (p. 5). That we are to understand this litany of complaints as the attribution of biological determinism becomes clear when Sahlins raises his objection to sociobiology. "The sociobiological reasoning from evolutionary phylogeny to social morphology is interrupted by cultures. . . . For between the basic drives attributable to human nature and the social structures of human culture there enters a crucial indeterminacy," so that "the same human motives appear in different cultural forms," and "psychological dispositions can take on a multitude of institutional realizations." Culture, Sahlins claims, "is the essential condition of this freedom from emotional or motivational necessity," for it enables man to interact in terms of sustained systems of meanings, with his emotions, and organize them (p. 11). He concludes that our emotional and motivational structures belie the claims of sociobiology just because they reflect a long course of cultural selection. But this conclusion is the very one Wilson embraces in his own discussion of human sociobiology.[15] Indeed, he says that in *Homo sapiens*, "the genes have given away most of their sovereignty, though they maintain a certain amount of influence in at least the behavioral qualities that underly variations between cultures."[16] Of course, Sahlins is right to reject the straw man of strong biological determinism, and Wilson is guilty of having misled his readers in his claims (obviously influential, though not explicitly cited by Sahlins) about ultimate and proximate causation in the determination of human behavior. But, as we have seen, although sociobiology is committed to what Sahlins calls the self-maximization of the gene, it is committed to little else he accuses it of. It is not committed to the treatment of human emotions and needs, under their current characterizations, as phenotypes; still less is it committed to a necessary correspondence between these states and conditions, however characterized, and human social relations. Surely Sahlins does not mean to deny the innocuous claim that social organization is the outcome of the "behavioral interaction of organisms having fixed biological inclinations." For on its most plausible reading this is simply the conjunction of the claims that humans have fixed inclinations—which, *broadly speaking*, is a truth of genetic theory—and that social phenomena are additive aggregations of individual behavior. To deny this latter claim is to dispute methodological individualism and not sociobiology. Sahlins' claim must therefore be understood as the attribution of the view that the sum of individual genetic inheritances is causally sufficient for all features of the character of social organization. But not only is this view one that Wilson repudiates, but it is also incompatible with the thesis of explanatory biological determinism that he requires. The sociobiologist is constrained to agree that culture does constitute a crucial indeterminacy in the causal chain from genes to institutions. But it is an indeterminacy that forecloses not just the sociobiologist's attempt to explain systematically as much of social structure as we want, but also everyone else's chances of doing so as well. Sahlins' view must be that the indeterminacy generated by the interposition of culture is relative: it prevents the discovery of systematic relations between genetically programmed drives and social structure, but does allow for the existence of acceptable explanations of such structure, autonomous from genetic inheritance. This is the real dispute between the sociobiologist and his opponent—the dispute over the acceptability of explanatory biological determinism. The anthropologist argues that access to the sustained system of emotion-controlling meanings, which both he and the sociobiologist agree cannot be assimilated either to evolutionary or neurological terms, can prove real explanations between the levels of details of these two theories. And of course, the sociobiologist denies this. The threat of preemption of his discipline which Sahlins recognizes, turns, as his own arguments suggest, on the dispute about this issue, and not on the disputes about the obviously absurd strong thesis of biological determinism or about the applicability of kin selection to the explanation of any particular sociologically individuated fact. But although he recognizes the threat of preemption, he only intermittently sees that it hinges on explanatory biological determinism and not on these other nonexistent or highly narrow issues. The same, of course, can be said for the sociobiologist who misleads himself and others into supposing that he is committed to an absurdity, and that his case will be made or broken on the issue of altruism.

To see this we need to turn to the next chapter and most sustained argument of Sahlins' book, "Critique of Sociobiology: Kin Selection." He begins by stating quite explicitly that the success of sociobiology turns on the fate of its theory of kin selection. For according to Sahlins, kinship is perhaps the most significant and ubiquitous institution in the primitive societies currently encountered, and therefore of inestimable importance throughout most of human history. And, Sahlins claims, sociobiology purports to provide a theory accounting for its importance and a theory explaining how kinship behavior is ordered. In particular, the explanation sociobiology offers for kinship is alleged to be a special instance of its general theory that reproductive success is the mainspring of social behavior. On the other hand, claims Sahlins, "if kinship is not ordered by reproductive success, and if kinship is central to human behavior, then the project of an encompassing sociobiology collapses." Thus, "the issue between sociobiology and social anthropology is decisively joined on the field of kinship" (p. 18). Let us for the moment accept this conditional as true, without inquiring into the status of its antecedents and consequent. According to Sahlins, sociobiology accords kinship the greatest importance, for it is only in this way that it can explain that hereditary predisposition to altruism which should long ago have been extirpated by selection. The solution to this latter problem is, of course, the theory of kin selection, which converts altruism into genetic egoism: "In the face of moral codes and kinship classifications that do not correspond to the rationality of genetic self-interest, sociobiology responds that the genes 'know,' whatever the structure of individual self-consciousness." But, says Sahlins, "the fact is that actual systems of kinship and concepts of heredity in human societies, though they never correspond to biological coefficients of relationship, are the true models of and for social action. These cultural determinations of near and distant make up the de facto form taken by shared interests and manifested in altruism and antagonism. They represent the effective structure of sociability. . . . There is not a single system of marriage, postmarital residence, family organization, interpersonal kinship, or common descent in human societies that does not set up a different calculus of relationship and social action than is indicated by the principles of kin selection" (pp. 25-26).

Sahlins purports to make good these claims and undercut the position attributed to sociobiology by first deploying a multitude of ethnographic findings on the variety of kinship systems observed throughout the world, and then by focusing on what he claims is a favorable case for the sociobiologists, if any is. He begins by noting that many primitive peoples operate in accordance with kinship rules that determine genealogical relationships through strictures on residence, either patri- or matri-local rules; these rules bring together genetically diverse individuals, and separate genetically close ones. Similarly, with patrilineal and matrilineal rules, such as those cited in the dispute between Needham and Homans described in Chapter 3, half the genes of a social interacting lineage are lost to that lineage in each generation, and half the genes are entirely new. Since such rules and the unions consequent on them separate the ties of and the occasions for altruism in a way that is utterly at variance with the genetic theory of kin selection, Sahlins insists, "the factors determining reproductive success are organized independently of genealogical relationships. . . . cultural order, as a symbolic and creative force, [is] not bound to express some natural kinship, but to invent kinship in the first place as a social form. Such an invention is clearly seen in this: that whether kinship is traced through two brothers or a brother and a sister constitutes a fundamental social difference, though it makes no genetic difference" (pp. 31-35). Sahlins instances as phenomena incompatible with the sociobiological "theory" of kinship the existence of ghost marriages, in which the dead are married to the living, and the issue produced by a surrogate who does not play the required altruistic role; or again, there are genealogies which accord females the role of father in the kinship structure making fatherhood nonsexual. Such cases show "that for human beings, survival is not measured in terms of life and death, or as the number of genes transmitted to future generations, humans do not perpetuate themselves as physical but as social beings" (p. 36). According to Sahlins, the cultural concepts underlying all apparent violations of natural selection "motivate a structure of human kinship that alone can account for the empirical form of an individual's social interests" (p. 39). He concludes his general survey by noting that members of the kinship groups that actually organize human reproduction are more closely related genealogically to persons *outside* their groups than to certain ones within them; accordingly, reproductive benefits are accorded in these groups to genetically unrelated persons; and thus, "discontinuities between the ethnographic topology of benefit relations and the natural structure of consanguinity generate irrationalities in the cost/benefit program alleged to control social behavior" (p. 22).

Not content with these arguments, Sahlins turns to an extended report on kinship in Polynesia, for it should provide, in his view, a particularly favorable case for the sociobiological account of kinship. Here descent is bilateral, post-marriage residence is optional, genealogies are preserved, and the indigenous theory of heredity corresponds more closely to biological fact than those of other peoples do. And yet these groups neither do nor can live in accordance with rules compatible with kin selection. For one thing the Polynesian would have trouble even expressing these rules, for his kin language is arithmetic, not geometric as kin selection's expression requires. Indeed, he cannot even express the fractions which this theory claims govern the distribution of altruistic acts that constitute kin groups. (As an aside [p. 45] Sahlins suggests this objection also undercuts the applicability of the theory to animals!) Like the others, the case of kinship among the Polynesians shows that no system of human relations is organized in accordance with the genetic coefficients of relationship known to the sociobiologist. Each consists in apparently arbitrary rules of marriage, residence, and descent, from which distinctive kinships derive, which violate kin-selection theory; and each reflects its own theory of heredity, which is never the theory of modern biology. Since these culturally constituted kinship relations govern the real processes of cooperation, property, aid, marital exchange, and so on, the human systems that order reproductive success reflect an entirely different calculus and demonstrate that human kinship is a unique characteristic of human societies distinguished precisely by its freedom from natural relationships: humans reproduce as social beings, and what are reproduced in human cultural orders are not biological organisms, but systems of social groups and the categories and relations in which they live. Culture is the indispensable condition of this system of social organization. At best, biology sets limits within which culture constructs a symbolic order, an order that can no more be reduced to a biological one than, Sahlins alleges, biological theory can be reduced to physical theory.

Much of this litany misses the point in dispute, as the following passage from Wilson, *quoted by Sahlins*, suggests: "The extent and formalization of kinship prevailing in almost all human societies are . . . unique features of the biology of our species."[17] Although this passage reads like an admission that sociobiology can tell us little about the details of kinship systems ever actually used or currently in use, since they are unique to our species, Sahlins quotes it to suggest agreement by sociobiologists about the centrality of an account of kinship to the dispute between him and Wilson. The passage clearly implies no such concession. Neither does Sahlins have grounds for claiming common ground in respect of other assumptions on which to erect his argument. Thus, the claim that reported systems of kinship among a few hundred currently existing societies represent facts of the highest significance in human evolution seems little more than the assertion that the special research object of the cultural anthropologist should have privileged status because of its ubiquity among his subjects. But no independent warrant is given for the alleged centrality. Moreover, the very heterogeneity that Sahlins reports among primitive societies' kinship systems suggests that the differences among them may be without selective significance at the present level of human evolution, and thus render unsurprising the sociobiologist's inability to explain them. On the other hand, greater homogeneity would play equally into the sociobiologist's explanatory hand. This, of course, reflects one of the features of any extremal theory: that it can make short work of the evidence whichever way it leads. But more important, not only need the sociobiologist not grant the importance of explaining particular kinship patterns, but the idea that his theory of kin selection should do the job reflects little more than a pun. Kin selection is a theory which explains the occurrence of undeniable hereditary properties of insects and certain species of birds, and which also renders formally consistent the hypothesis of natural selection of the fittest with the existence of "altruistic" traits among the members of any species. "Kin" as it figures in this theory refers only to individuals sharing a significant subset of their genes in common. Since it is obvious that as "kin" figures in the marital prescriptions and proscriptions it is not defined in terms of gene-sharing, it follows that any coincidence between these two terms may be just that: coincidental. Since the theory of kin selection explains the origin and persistence of altruistic traits and acts at relatively early stages of the development of a species, it is no surprise that when selective pressures change, other forces, both hereditary and environmental, may come into play and generate self-imposed kinship terminologies and rules bearing no connection to those forces of kin selection which, perhaps initially, selected favorably particular groups. Sociobiology offers *no direct* explanation for ethnographic kinship; the claim that it does so as a special case of its theory of kin selection is false and serves simply to enable Sahlins to erect a straw man which he can render speechless with ethnographic minutiae just because it has nothing to say on the subject.

Sahlins does not mention that many of the societies on which anthropologists have reported are so small that all their members share much greater than average numbers of genes in common, so that evolution away from common behavior in accordance with kin-selection theory is just what sociobiologists might expect. He does not mention that species in which genes are exchanged among larger numbers of members generally are fitter because of hybridization than those in which they are not; so that kinship rules which encourage the marital dispersion of members of a lineage may reflect the operation of selection even more strongly than kinship systems which favor only genetically close members. Nor does Sahlins mention that the most influential anthropologist among those who have studied kinship systems seems to favor the very evolutionary explanation of the character of kinship systems that Sahlins rejects: as we saw in Chapter 3, Lévi-Strauss explains the prevalence of one sort of cross-cousin marriage over another by according societies employing it a greater organic solidarity that provided them an evolutionary advantage over societies with other sorts of rules. Finally, Sahlins ignores two other mechanisms which the sociobiologist employs to explain group behavior and social institutions among a variety of species: group selection and reciprocal altruism. Indeed, Wilson quite explicitly commits himself to the view that reciprocal altruism, a phenomenon he does not consider hereditary, is a more likely source than kin selection of the sorts of social institutions prevalent among human societies today.[18]

But all this is to grant that there is even an issue between Sahlins and the sociobiologists on the nature and origin of kinship systems. The fact is that the real issue is at the level of generality of the thesis of explanatory biological determinism. Thus, if Sahlins is correct that none of the groups studied by anthropologists organizes itself in accordance with genetic coefficients of relationship, and that each employs arbitrary rules and its own idiosyncratic theory of heredity, then this just gives further reason to ignore the groups' claims in the explanation of their behavior, or to conclude that no systematic explanation of their behavior at the level the anthropologist hopes for is possible. To claim that kinship is a unique characteristic of human societies is not thereby to insure the autonomy of the science which studies it, but to admit the impossibility of a scientifically respectable account of kinship; it is to treat the behavior described by such rules as reflecting at best accidental generalizations, which should be no surprise considering the frequency with which they are breached. In short, all the evidence that Sahlins mounts to buttress the autonomy of culture from biology is but more evidence for the sociobiologist's claim to preempt the anthropologist's territory. For the latter, in effect, announces that his findings are expressed in a language not connectable with the rest of the language of science, and that we cannot expect generalizations to emerge from his study which we can confidently employ in the explanation and prediction of human behavior. Sahlins is correct to claim that culture generates a crucial indeterminacy between basic drives and social structure. But this indeterminacy haunts him more then the sociobiologist. For in consigning his own study of such structures to what he calls the system of sustained meanings which control and organize our emotions, he effectively excludes himself from anything more than an anecdotal collection of singular statements without systematic import. This, at any rate, will be the consistent sociobiologist's response to Sahlins' arguments. He will accept them as confirming his own claims about the need to supplant subjects like anthropology—which, after all, is but a narrow branch of ethology—by a more powerful science. To the extent, however, that the sociobiologist accepts the terms of Sahlins' discussion and attempts to employ or defend the employment of a general theoretical explanation of the possibility of a phenomenon he calls "altruism," in spite of its patent dissimilarities to the form of morally supererogatory behavior that ordinarily goes by that name, to that extent he renders his theory vulnerable to complaints of hollowness and tautology, or falsity and irrelevance, and permits the social scientist to hold out the possibility of a science between the levels of Darwinian theory and neurophysiology. This mistake is what has lent the debate over altruism an importance all out of proportion to its theoretical role in the panoply of sociobiological models.

It is worth noticing an important similarity here between the views of the two self-styled defenders of the autonomy of anthropology that we have examined in this work, Sahlins and Needham. The writings of both share a mixture of bemusement at and contempt for anyone who would venture hypotheses about anthropologically significant phenomena without first making extensive field expeditions. More significantly, both manifest an intense desire to insulate their own study from any other, and to assure its total substantive and methodological autonomy no matter what the price in sterility and insignificance. Taken together, their arguments effectively cut all connections which we might normally suppose to obtain with subjects that treat human behavior at lower and higher levels of aggregation than are treated by anthropology. Thus, Needham's aim in *Structure and Sentiment* is to establish the claim that "whenever a social phenomenon [like kinship in particular] is directly explained by a psychological phenomenon, we may be sure that the explanation is false,"[19] and Sahlins' aim is to sustain the conclusion that the same is true for biological attempts to explain the very same sort of social phenomenon. Both anthropologists argue for their claims by citing anthropological findings which they allege to disconfirm single psychological or sociobiological theories of kinship. Naturally, their conclusions therefore transcend their evidence, and in fact rest on much more than the evidence deployed. The degree of insulation which they seek and find for their discipline could be based only on a philosophical commitment incompatible with the eventual assimilation of their subject to others. A commitment to the emergence of culture as an entity different in kind from those which other subjects treat is essential to their view, and also renders discussion of their differences with empiricists pointless. Having foreclosed their account of kinship either to reduction to psychological dispositions, or subsumption under selective forces, they have effectively and perhaps purposely cut their subject off from aid or assault in any direction. It would be an interesting sociological question why anthropologists have felt the need more strongly than other social scientists to mount arguments for the autonomy of their subject, and to reject with a show of the greatest contempt the pretensions of anyone but a member of their tight little circle to employ, let alone explain their alleged findings. That is, it would be an interesting question, if the philosophy of social science here expounded permitted the provision of systematically significant explanations of social events at so low a level of specificity. On the account of the possibility of social science sketched here, any statement of why anthropologists are sensitive about matters of methodological and substantive autonomy of their subject would be at best a true singular causal statement; and since there are no laws about the causes of anthropologists' sensitivities, or about any other states of *Homo sapiens*, or students of *Homo sapiens*, described as such, the singular statement even if true would be without the systematic backing required to figure as a scientific explanandum.

In a way the fact that this question seems so natural a one to ask reflects the most serious problem for the strategy here employed to render consistent empiricism with some of our ordinary beliefs and with the evident nomological failure of social science. Not only are we disposed to ask such questions as social scientists, we are even more strongly disposed to do so as normal persons. It would be foolhardy to suppose that reading this book will in fact cause anyone to stop citing beliefs and desires as the causes of human events for which explanations are sought. What is worse for my purposes, it has been obvious throughout this argument that despite my attempt to avoid the terms 'belief' or 'desire' in explaining why particular views are held, I have nevertheless done so by citing such reasons. And the fact that I have written a book suggests that I hope to change your beliefs, and perhaps some of your actions—among them, in particular, your speech acts—by changing others of your beliefs, assuming that you have certain desires, held constant. What is more, both within this work and beyond it, I and everyone else assent to subjunctive conditional statements and counterfactual ones about how individuals would behave if their beliefs and desires were different. My actions and everyone else's seem to belie my words. The way technically to put this point is to say that we clearly *do not treat L*, our exceptionless general statement about reasons and actions, as an accidental generalization. Not only is it incumbent on me to explain why we are willing to embrace counterfactuals that rest on L, if L is merely accidental, but also to explain the intelligibility of the behavior involved in writing a book, arguing for a conclusion, attempting to convince readers, asserting propositions, condemning arguments, when the very characterization of the behavior that makes it intelligible rests on embracing the conceptual scheme and its associated general statement that I have spent all these pages repudiating as the acceptable foundations for a systematic explanation of anything. Notice, to my embarrassment, that the anthropologist has no difficulty with this question, and can explain the intelligibility of my behavior in a way that is of a piece with the explanation of the behavior of the most exotic and distant tribe. Needham and Sahlins can point out that just as the kinship rules governing the behavior of members of a group are embedded in the symbolic system of relationships reflected in the language of this group, so too are the rules making my behavior intelligible action embedded in our language. This is why the anthropologists must go native: not merely to learn how to communicate with the informant, but because going native, learning the language, is already to provide oneself with the material that makes the native's life intelligible. Among these rules in our own society are those which make my own behavior intelligible as actions, and which make sensible the description of my behavior as that of writing a book, asserting propositions, and attempting to convince others, to change their minds, to affect their research programs. But if the aim of my book is to deny the explanatory power of the network of concepts which are employed to describe its aim, then the whole project is self-refuting. Or so it might be alleged. The use of language presupposes the intelligible applicability to events of categories like belief, desire, intention, and action. To deny as I have the explanatory standing of this whole circle of concepts is to destroy the very tool required to effect my denial. Surely I will agree that if I had not had the beliefs reflected in these pages, or the desire to lay them before others, I would not have written this book. But if I agree to this counterfactual claim I must, it is argued, assent to the general statement, like L, which sustains it. To do this is to admit that my actions, the production of this book, undercut my words, my denial that L is anything more than an accidental generalization. For it is a hallmark of such statements that they lack all counterfactual force. If it is logically or pragmatically impossible for me to express coherently the views here inscribed, in the way that it is impossible for me to announce coherently and sincerely "I am not here now," then these views have no more claim to attention than we have power to advance or embrace them.

This point could equally well be made by an anthropologist like Sahlins or Needham, or by a philosopher like Winch. Thus, Needham writes that understanding human behavior "involves a systematic comprehension of the life of a society in terms of the system of classification employed by the people themselves."[20] And Sahlins explains that the system of classification—that is, language—is the core of the "reason why human social behavior is not organized by the individual maximization of genetic interest," nor by the neurophysiology of human agents for that matter. For "human beings are not socially defined in terms of their organic qualities but in terms of symbolic attributes; and a symbol is precisely a meaningful value" (p. 61). The importance of language, in the anthropologist's view, lies not in its "function of communication," but in its "structure of signification," which confers on speech the "creative action of constructing a human world: that is by the sedimentation of meaningful values on 'objective' differences according to local schemes of significance. So far as its . . . meaning is concerned, a word is not simply referable to external stimuli but first of all to its place in the system of language and culture. . . . By its contrast with [other words] there is constructed its own valuation of the object, and the totality of such valuations is a cultural constitution of 'reality' " (p. 62). We may render this evocation of the power of language both to reflect and to rule our lives in more pedestrian terms, by noting that the anthropologist, and the philosophers who agree with him about the role of language as a repository of our notions about "reality," claim that an attempt to deny a place to notions as central to our conceptual scheme as reason and action in the explanation of our own behavior must be false, for the use of language presupposes the intelligible and the correct description of our behavior in the very terms whose applicability we seek to deny. In order to understand the behavior of an alien people, we must learn to describe their behavior in their own terms, for only in these terms will action be distinguishable from reflex, habit from ritual, affirmation from negation, belief from desire, and belief and desire from fear or envy. Similarly, in order to understand, express, accept, or reject an argument to the effect that one scheme of explanatory concepts should be preempted by another, we must be able to describe coherently and instance cases of understanding, expressing, accepting, and rejecting arguments. But the scheme of concepts whose preemption I urge on behalf of sociobiology and empiricism are the very ones we require in order to describe my aims. In a way, therefore, my argument is self-refuting, for its expression entails the "reality" of just what it denies reality to. Or so it may be claimed.

I have said that this point could be made equally by an anthropologist eager to insulate his subject against preemption, or by a philosopher, like Winch,[21] eager to assimilate all social science to a species of philosophy of language. Perhaps the best and clearest expression of this line of argument, though without the explicit aims of either Sahlins and Needham, or Winch, is due to Norman Malcolm.[22] Malcolm advances the claim as an argument against mechanism, which he defines as the thesis that there is a neurophysiological theory, as yet undiscovered, "which is adequate to explain and predict all movements of human bodies except those caused by outside forces" (p. 45). Clearly this is a thesis with which the physicalist and the sociobiologist are, to say the least, in sympathy. Indeed, the sociobiologist accords his own evolutionary theory acceptability only because it is a theory ultimately reducible in part to the neurophysiological theory that mechanism requires. The sociobiologist is a mechanist. According to Malcolm, mechanism is not a conceivable doctrine, for at least two reasons: "The first is that the occurrence of an act of asserting mechanism is inconsistent with mechanism's being true. The second is that the asserting of mechanism implies that the one who makes the assertion cannot be making it on rational grounds." For

> if mechanism is true . . . we need to junk all such terms as "intentionally," "unintentionally," "purposely," "by mistake," "deliberately," "accidentally," and so on. The classifying of utterances such as "asserting," "repeating," "quoting," . . . "translating," would have to be abandoned. We should need an entirely new repertoire of descriptions of a sort that would be compatible with the viewpoint of mechanism. . . . If mechanism is true not only should we give up speaking of "asserting," but also . . . of "speaking." It would not even be right to say that a person *meant* something by the noise that came from him. No marks or signs would mean anything. There could not be language. . . . If [the mechanist] is right we do not use concepts at all. . . . The motto of the mechanist ought to be: One cannot speak, therefore one must be silent (p. 71).

It is essential for our purposes to refute the argument that leads Malcolm to this conclusion. For surely if his argument is sound, the conclusion completely undercuts the program here advanced for social science, and the solution to the trilemma for empiricism here proposed. This is especially true because many of the crucial premises of my argument are ones Malcolm seems to employ in his. Moreover, while it is easy to fob off the exaggerated protestations of anthropologists against the denial of the existence of their subject, Malcolm offers a literally less embellished, and far more precise, version of their own argument. It may be expected that the neurophysiologist and the sociobiologist will not allow themselves to be detained by what they consider at best Malcolm's philosophical sleight of hand, but this will only reveal the shallowness of their own appreciations of their subjects' presuppositions. Finally, successfully refuting Malcolm will enable us to reemphasize some of the features of the argument of this book and to explain the role of the true singular causal statements we make about human behavior every day.

Ex hypothesi, the mechanist's hoped-for neurophysiological theory will cite the neurophysiological conditions which are causally sufficient for each and every internal movement of a human being, presumably by appeal to laws that are not restricted in their applicability to *Homo sapiens*. Thus, for any event we might describe as an action and explain by citing its reason, the neurophysiological theory provides the resources to construct an alternative explanation of the same movement that makes no mention of desires, aims, goals, motives, intentions, beliefs, hopes, or fears. Malcolm rightly insists that the explanations that do cite these latter factors implicitly or explicitly appeal to general statements just as neurophysiological explanations do. And these, in fact, have forms which make them substitution instances of L, the exceptionless statement connecting beliefs, desires, and abilities with actions. Malcolm treats these principles as a priori truths, whereas we have not. But for our immediate purposes this disagreement is not material, especially since Malcolm believes that a priori truths like these can provide causal explanations of the occurrence of particular events. Now, Malcolm asks, what is "the exact logical relationship between neural and purposive explanations of behavior[?] Can explanations of both types be true of the same bit of behavior on one and the same occasion? Is there any rivalry between them? . . . Do the two accounts interfere with one another?" (pp. 51-52). Malcolm invites us to consider the example of a man climbing a ladder up to a roof on which lies his hat, it having been blown there by the wind. Since the neurophysiological theory provides the causally sufficient conditions of his behavior, "the movements of the man on the ladder would be completely accounted for in terms of electrical, chemical, and mechanical processes of the body. This would surely imply that his desire or intention to retrieve his hat had nothing to do with his movement up the ladder. . . . He would have moved as he did regardless of his desire or intention" (p. 53). Thus, "there would be a collision between the two accounts if they were offered as explanations of one and the same occurrence of a man's climbing a ladder" (p. 52). Of course, there would be no collision, as Malcolm recognizes, if what he calls "the currently popular psychophysical identity thesis" were true; that is, if "there is a neural condition that causes the man's movements up the ladder, . . . and the man's intention to climb the ladder is contingently identical with the neural condition that causes the movements" (p. 53). Malcolm raises a number of objections to this consequence of the identity theory, no one of which has been treated as very serious by its proponents. But we must accept Malcolm's objection, for we have in effect already admitted that the property of having intentions, or even of having a particular intention individuated by reference to its propositional content, is not identical to any manageable small disjunction of neurophysiological states nor associated with any topographically specified behavior with the kind of regularity required for its independent specification. The claim, repeatedly made, that notions like intention are not natural-kind terms is tantamount to the admission that no claim of identity, with nomological force, between mental states and brain states is forthcoming. But this is just what the exponent of the psychophysical identity theory requires. So quite regardless of whether the materialist finds Malcolm's argument uncompelling because the latter has not excluded the identity theory to its proponent's satisfaction, we cannot take leave of Malcolm's argument at this point. We must agree that a neurophysiological account of the event in question and an explanation of it that appeals to L must collide, must be rivals, must interfere with each other, cannot both be true of the same bit of behavior at one and the same time.

"Thus if mechanism is true, . . . principles of action [like L] do not apply to the world. This would have to mean one or the other of two alternatives. The first would be that people do not have intentions, purposes or desires, or that they do not have beliefs as to what behavior is required for the fulfillment of their desires and purposes. The second alternative would be that although they have intentions, beliefs, and so forth, there are always countervailing factors—that is factors that interfere with the operation of intentions, desires and decisions" (p. 63). Again, Malcolm excludes the second alternative through means that might be construed as controversial by some exponents of mechanism, but which our previous argument forces us to accept. He claims that "it is not a coherent position to hold that some creatures have purposes and so forth, yet that these have no effects on their behavior. Purposes are, in concept, so closely tied to behavioral effects that the total absence of behavioral effects would mean the total absence of purposes and intentions. Thus the only position open to the exponent of mechanism is the first alternative—namely, that people do not have intentions, purposes, or beliefs" (p. 64). Malcolm's justification here is the allegation that statements like L are a priori. Although we do not embrace this view, it has been part of our argument that L provides the sole available means for identifying desires, beliefs, and actions, and thus plays the sort of role that leads Malcolm to construe his versions of L as a priori. We cannot follow him in this because of empiricist strictures that preclude the employment of a priori statements to effect the nomological connection in causal explanations. Malcolm seems to have no such scruple, but we cannot reject his argument on that account, for we can supply alternative premises.

Thus Malcolm concludes, "A mechanist must hold . . . that . . . principles of action [like L] have no application to reality, in the sense that no one has intentions or desires or beliefs. . . . Mechanism is incompatible with the existence of any intentional behavior. The speech of human beings is, for the most part, intentional behavior. In particular, stating, asserting, or saying that so and so is true requires the intentional uttering of some sentences. If mechanism is true, therefore, . . . no one can assert or state that mechanism is true. If anyone were to assert this, the occurrence of his intentional 'speech act' would imply that mechanism is false" (pp. 64-67). There is a second ground for the inconceivability of mechanism besides the conclusion that it is not stateable: "Saying or doing something *for a reason* (in the sense of grounds . . .) implies that the saying or doing was intentional. Since mechanism is incompatible with the intentionality of behavior, my acceptance of mechanism as true for myself would imply that I am incapable of saying or doing anything for a reason. . . . [Thus] not only would the assertion [of mechanism] be inconsistent, . . . but it would also imply that I am incapable of having rational grounds for asserting anything, including mechanism" (p. 70).

Unlike most arguments against mechanism based on the primacy of intentional descriptions of behavior, this argument seems especially compelling. It does not beg the question by presupposing the applicability of notions like belief, desire, and action to the correct description and explanation of human behavior. It makes no appeal to paradigm cases of the sort reflected in our conviction that at least some singular statements about particular reasons and their effects in specific actions must be true. It gives the thesis of mechanism as much a priori possibility as that thesis's proponents could ask, and unlike many defenses of the primacy of intentional language, it recognizes that neurophysiological accounts are to collide with accounts based on principles like L, and to supplant them, as a thesis of explanatory biological determinism "intends." Moreover, it is more powerful than an argument which simply shows that we must surrender intentional descriptions and explanations if mechanism is true. As Malcolm notes, such arguments may be blunted by the attitude that "this consequence cannot refute mechanism or jeopardize its status as a scientific theory. It would seem to be up to science alone to determine whether or not there is a comprehensive neurophysiological theory to explain all bodily movements in accordance with universal laws. If scientific investigation should confirm such a theory, then so be it! . . . If confirming the theory were to prove that people do not have desires, purposes, or goals, then this result would have to be swallowed, no matter how upsetting it would be not only to our ordinary beliefs but also to our ordinary concepts" (p. 69). Clearly, this line, no matter how likely to find its way to the lips of defenders of mechanism, will be of no avail, for Malcolm's argument shows much more than the incompatibility admitted by the defense. It purports to show the unstateability and the rational unbelievability of the thesis defended by this line of argument. This is what makes the argument so compelling.

How can we circumvent Malcolm's conclusion? We cannot give up the possibility of the neurophysiological theory that mechanism requires. Even if we could give up our commitment to the truth of singular statements about particular reasons and actions, it would be of no avail in refuting Malcolm. The controversial premises of his argument that deny the truth of the psychophysical identity theory and the apriority of principles like L may be rejected by others, but our previous arguments have provided enough grounds for Malcolm to construct new premises that do the work the unacceptable original premises do. The problem that faces us reflects so many features of the strategy of this book that solving it, circumventing Malcolm's conclusion, would itself constitute a powerful consideration in favor of the plausibility of the position here defended; equally, failing to do so provides enough reason to reject it as incoherent and intrinsically unjustifiable. For consider, the assumption of empiricism and physicalism made in the second chapter entails the possibility of the mechanist's neurophysiological theory. The analysis of social scientists' self-conscious attempts to avoid appeal to intentional variables in the explanation offered in the third chapter showed how difficult it is to consistently surrender this language, or to deny its causal force. The fourth chapter offered a chronicle of the failures of social scientists to provide systematic behavioral accounts of the notions of desire, belief, and action, and the fifth undercut the possibility of providing neurological correlates for them. In effect, these chapters represent sustained arguments for the premises of Malcolm's *reductio ad absurdum* of mechanism. The sixth and seventh chapters commit the empiricist and the physicalist to the employment of biological theory in the explanation of human behavior, and commit him as well to a preemption of conventional social science by sociobiological research strategies. In effect, these latter chapters reflect the collision between nonintentional and intentional explanations of behavior that Malcolm also requires. What resources do these chapter provide for refuting, and not sustaining, the inconceivability of the very position for which they argue?

All the resources we require for circumventing Malcolm's conclusions are available in the separation effected in Chapters 5 and 6 between the referential employment of the notions of desire, belief, and action, on the one hand, and their explanatory use on the other. This distinction, in turn, rests on the existence of the states, conditions, and events to which the antecedents and consequents of *L refer*, and the spatiotemporally restricted character of the properties which these antecedents and consequents *attribute* to the states, conditions, and events to which they refer. Since the properties attributed are not nomologically permissible ones because they make tacit reference to a spatiotemporally restricted particular object—the species *Homo sapiens*—it follows that any general statement connecting their instantiations to one another must either be a definition, the consequence of one or more definitions, or an accidental generalization. As such, the general statement cannot be employed to explain the consequences of a state, condition, or event *successfully* referred to by a singular statement employing terms describing beliefs or desires. Nor can such a statement explain the conditions of an event again successfully referred to by a statement employing terms that describe the event as an action of any kind. The descriptive employment of reason and action terms is often successful. That is, statements using these terms often secure reference to the states, conditions, and events, in which humans figure, that make them *true*. And sometimes statements securing reference to conditions of a person that make them true are linked in causal contexts to other statements about the movements of humans that make true statements successfully referring to these movements. Under these circumstances the entire causal statement will be true. But the truth of causal statements will not entail that the properties attributed in their cause and effect clauses are the properties by virtue of whose manifestation the whole causal statement is true. The use of a certain body of descriptive terms to identify explananda events and explanans conditions, does not entail that these terms will figure in nomological explanans statements or the systematically general description of the explanandum event. We may accept all the singular statements about particular reasons and their consequent actions as true without committing ourselves to the explanatory relevance of the terms we hit on to make these true statements. In particular, it is true that Napoleon ordered in the Old Guard at Waterloo because he believed that Blücher had already been defeated, for the first clause is made true by a particular physical movement made by a particular human body at a particular instant in the Belgian countryside near Brussels in the spring of 1815, and the second is made true by that body's being in a particular highly complex neurological and physiological state on and just before that occasion. Similarly, anyone who utters the noise that expressed the statement to this effect engages in a movement within and at the surface of his body which constitutes the utterance of that particular statement; and the cause of this movement is a neurological state of the same body which is successfully referred to as the belief that Napoleon did so for the reason given. But neither of these truths requires for its truth the truth of any general statement either about beliefs and actions or beliefs and speech-acts, in particular, or about beliefs about Blücher and dispatches of the Old Guard, or beliefs about Napoleon and assertions about his reasons for actions. Nor will these true singular causal statements require for their explanation generalizations like L, or its substitution instances. Indeed, if our analysis is correct, they cannot be explained by such accidental generalizations.

We have denied the existence of nomological identities between kinds of neurophysiological states and kinds of mental states, and this has enabled, nay, required us to deny that there are laws connecting the mental states with one another, and with behavior, qua mental states and actions. Thus, we have had to accept Malcolm's premised denial of psychophysical identity; but we have embraced, on the physicalist principles that lead to mechanism, the identity of particular mental states, conditions, and events of particular humans at particular places and times with some one or other of their neurophysiological and/or physical states at those same places and times. This enables us to admit that although couched in the spatiotemporally restricted vocabulary of intention and action, our singular causal statements of this kind can be true. Thus, any obviously plausible statement about the causes of particular movements described as actions, or the effects of particular physical states described as psychological ones, which Malcolm hopes to force the mechanist to deny himself and others is permitted as possibly true on our version of mechanism. Thus, I may legitimately claim to be asserting the thesis of mechanism, and the actual neurophysiological causes of my assertion that this thesis is correct can be successfully referred to by characterizing them as occurrent states of belief that mechanism is true and states of desire that others come to believe it. And others may, for all I know, come to manifest those neurophysiological states that we refer to as the belief that Rosenberg is correct about these matters. What we cannot do on the basis of my account is explain either my assertion or your state of belief by general statements couched in the language of belief, desire, and action. For, in the collision between mechanism and purposive principles, these must be surrendered, even though we may continue to use the language of these principles to make true if nomologically unsubstantiated causal claims.

We may agree with Malcolm that explanations of action that cite beliefs and desires—as opposed to causal statements that merely correctly secure reference to their causes—do require explanatory principles like L, but we insist that all such explanations are false because there is no law of the sort required. And we must remember that, given the spatiotemporally restricted character of notions like reason and action, our true singular statements have all the systematic content of a statement like "Fido caused Rover," where "Fido" and "Rover" are names we use to refer, respectively, to events like, for example, the Titanic's striking an iceberg, and its sinking, which are causally connected. "Fido caused Rover" is true, but is no explanation of why Rover happened. Similarly for Napoleon's belief and his action, as we ordinarily describe them. We may agree with Malcolm that if mechanism is true, "principles of action [like L] do not apply to the world." But we need not accept either of the two alternatives he infers from this; it will not be the case that people have no intentions, or desires, or beliefs about what action is required for their fulfillment. They may have all the states and conditions described in these words that we wish to accord them. Nor will it be the case that although people have these states which we thus describe, these states are always prevented from eventuating in movements, which we describe as actions. What will be true is that it is never *because* they are beliefs, or desires, that they result in the movements in question. We are not committed to the incoherent position that "some creatures have purposes, but these purposes have no effects on their behavior."

Each and every one of the singular judgements about intentional behavior and its intentional determinants that we deem to be true, and that are true (these two classes probably are not coincident), is true by virtue of the operation of a large number of neurophysiological, physiological, anatomical, and biochemical general laws, so many such laws, no doubt, that we shall not soon discover many of them. Even if we do they will be of little use in explaining, predicting, and controlling behavior to any topographically accurate degree. Nevertheless, it will be the satisfaction of the antecedents and consequents of these laws by the states and movements described in our intentional language that will make the singular statements true, or so the mechanist alleges.

Malcolm's argument for the inconceivability of mechanism is thus circumvented. But its plausibility and its general line of argument help us show why no one has recognized a proposition like L to be an accidental generalization; it helps us show why people mistakenly advance counterfactual statements about reasons and actions, which they sustain by appeals to one or another version of a principle like L, not recognizing its accidentality. The distinction between the referential use of language and its attributive use is a subtle one, so subtle that it is the product of only very recent work in the philosophy of language;[23] indeed, it is in many respects a controversial distinction. It is hardly to be expected that this distinction should work its way into the ordinary person's view of the nature of his own referential and attributive claims. On the other hand, the need for a justificatory principle to stand behind what we believe to be true statements about reasons and their effects is apparent to most reflective persons. Moreover, there is a vague recognition—evinced in the works of anthropologists, for example—that the language of belief, desire, and action is essential to our view of ourselves. For it is fundamental to the human race's conception of its members that we treat one another as *agents*, and this requires us to accord one another the intentional states in question. In all societies systems of ethics, mores, and rules of conduct have evolved, for causes that perhaps the sociobiologist can sketch. These systems presuppose the treatment of people as agents. The treatment accorded other species reflects our belief that they are not agents. The depth of this conviction is reflected in the age-long problem of free will versus determinism, in the recurrent appeal to retribution as a justification for imprisonment, in the occurrence and expression of sentiments of approval or disapproval for the behavior of ourselves and others and for our "creative" products. So much of our lives are conceptualized in terms that demand the applicability of beliefs, desires, and actions that it is impossible for us, even for a brief time, to suspend the belief that much human behavior is action, and is caused by reasons. The snares and apparent absurdities into which so suspending our beliefs easily leads are illustrated forcefully in Malcolm's very argument against mechanism. We are constrained to believe either that mechanism is consistent with the conception of ourselves as agents and not merely clockworks, or that it is false. From this unshakeable belief to the mistaken judgement that agency requires the truth of laws about *agents* and their states, as natural kinds, is a natural result of millennia of reflection. Our belief that statements employing such terms are explanations or first approximations to explanations of our behavior, that they provide means for controlling or assessing our behavior, commits us to treating general statements which seem formally to underwrite these explanations as nonaccidental. To treat one another as agents is to be willing to offer counterfactual statements about our behavior in terms of its intentional determinants. Since we do this, we are prevented from noticing that the generalizations to which we implicitly or explicitly appeal in order to sustain these counterfactuals are accidental ones.

Once it is admitted that our confidence in the generalizations is dependent on our felt conviction that the singular statements they summarize are true, instead of the other way around, as in the case of scientific laws, they are deprived of their role in the justification of the counterfactual assertions we offer about reasons and actions. But if we do not notice that this state of affairs deprives the generalizations we actually formulate of their nomological force, we continue to treat them as if they had such force. We fail to recognize that they are accidental. What is more, their centrality to our view of ourselves as agents makes us so psychologically incapable of surrendering them that we are often inclined to view them as Malcolm does, as a priori principles, as the consequences of definitions of their terms. As such we forgo the opportunity to employ them in empirical explanations of human behavior. And this, of course, is what leads many philosophers of social science to deny that the aims of these subjects are similar to those of natural science. For if these subjects seek to express, catalogue, and interconnect definitions and consequences of definitions, then they cannot hope to provide explanations or predictions of the occurrence of actual events, including those which instantiate the properties whose definitions they express, catalogue, and interconnect. Accordingly, the aims of social science, in these philosophers' views, are not those of natural science, and the empiricist's attempt to reveal or prescribe empirical methods in or for these subjects rests on a conceptual mistake about these subjects.[24] The fact that empirically motivated social scientists have been unable, over a long course of investigation, to produce exceptionless general statements about reasons and actions which can be entrenched in a nomological network, coupled with the apparent need for universal statements generated by our commitment to human agency and to moral and esthetic evaluation, all conspire to make this view plausible. But if the aims of a science of behavior *do not include justifying* as correct our commitment to the treatment of one another as agents who act in accordance with principles about our reasons and their consequences for action, *do not include justifying* as true or correct our evaluative beliefs about ourselves, our behavior, and our creations, then plausible or not, this argument cannot convince us that social science requires the sorts of principles here abjured. But surely, social science does not have among its duties the justification of these beliefs. It does have among its obligations the explanation of why we hold these beliefs about agency, about the rightness and wrongness of actions (as opposed to events), about the goodness or wickedness of persons, and their motives, about the beauty and ugliness of our intentional creations. That is, it has these explanatory obligations, provided they can be fulfilled. I have argued that within practical constraints they cannot be, for their only chance of explanation lies with the development of neurophysiological theory; and such a development will not eventuate in practically useful explanatory and predictive resources with respect to everyday, normal behavior, or its aggregation. But although I hold that we cannot explain action by appeal to belief and desire, and that these notions have no place in a scientific explanation or description of behavior, I do not deny that there are beliefs, desires, and actions; that sometimes cases of the first two cause cases of the last; and that sometimes we are *correct* in believing that this has occurred. I do hold that although we are sometimes correct in our singular causal judgements about actions and their determinants, we are *not* justified, *by scientific standards of justification*, in holding these beliefs, for we do not have the requisite laws in hand to justify them. We may well be justified in holding these singular beliefs in the light of other, nonscientific standards. But if the empiricist is correct, these standards will not be epistemic ones, for the only epistemic standards are the ones reflected in science. They may be standards drawn from the canons of law or a system of ethics or abstracted from the exigencies of practical life. But to this extent these standards will reflect general statements that are at best only accidentally true, and at worst exception-ridden or empty of content.

The claim that none of our singular causal statements about particular reasons and actions (including the true ones) is justified must be distinguished from the claim that the denial of each of them (again including the true ones) *is justified*. And while we are committed to the former claim, we are certainly not committed to the latter. This latter claim is, of course, the one which Malcolm hopes to pin on the mechanist, and the one that most reflective people suppose is the alternative to the justification of our beliefs about actions and their causes. It is this latter claim that seems incompatible with the admission that at least some of these judgements are true. For it is surely paradoxical to believe that a proposition is true, and that its denial is justified. On the other hand, to admit that none of a class of judgements is justified, even though one or more is true, is a perfectly possible state of affairs. This is the state of affairs our theory requires. And the fact that none of the singular judgements we make about other people's beliefs, desires, and consequent actions is immune to withdrawal or revision reflects the fact that their justification is at best incomplete. Indeed, if our judgements rested only on our access to the behavior of others, and not on access to our own cases, we should long ago have surrendered the search for laws of human action. Most of the strengths of conviction that at least some such statements are true derive from introspection, from reflection on our own behavior and its conscious correlates. But although this introspection accords confidence to the truth of these statements about the specific reasons that determine our own particular actions, they accord no confidence to the truth of any general principle that will turn these causal truths into causal explanations and provide materials to predict our own and others' behavior. The oft-noted paradox that we can decide on our own behavior but in an important sense cannot predict our own behavior reflects this fact about the findings of introspection and their relevance to these matters. "When I look most intimately into what I call myself," as David Hume expressed it, I find the operation of no principle like L or its alternatives in "operation," and if asked why my reason led to my action, I can only say, "Because it was my reason," or more generally, but no more helpfully, I may shift from the first person perspective and announce that in creatures like me states like these lead to actions of the kind discharged. While introspection may assure us that some of our singular statements of this sort must be right, the philosophical and moral problem of *Akrasia* that has attracted attention from Socrates' time to our own is testimony to our willingness to suspend confidence in the truth of any given singular statement, and to admit that merely finding the reasons for an action may be no explanation for it. The problem of Akrasia, or weakness of the will, is that of explaining how it is possible for someone to fail to do what he wants to do and is able to do. But the question of how this is possible presupposes that it is possible, and therefore that having the appropriate wants, abilities, beliefs, and knowledge does not necessitate causally, logically, or in any other way that the act will follow. Entertaining this possibility is tantamount to denying the existence of laws of human action and rendering unjustified all the singular claims to the effect that a particular set of beliefs, and desires, in the context of certain abilities, and in the absence of any impediments, caused the actions that follow them.

Everything I have said in the last seven paragraphs about the causes and effects of various beliefs about matters practical, logical, and ethical, must be construed as a body of singular judgements whose justification lies in the conviction I share with everyone else that some such statements about the causes and effects of mental states and actions are true, *and* the conviction I share with empiricists and physicalists, but not with others, that one version or another of these two philosophical doctrines is true. Without these theories, the beliefs I have expressed in these paragraphs—and indeed, in the whole work—are no more justified than the beliefs of ordinary people and social scientists that their favorite singular statements about reasons and actions are true. Without the support of these doctrines, I would indeed be hoist by my own petard, as Malcolm hopes to show. For I would be guilty of offering the reader a vast body of my beliefs about causal connections among mental states and between such states and behavior, that, on my own terms, was unjustified, and therefore without recommendation to anyone's attention. Accordingly, anyone who rejects these philosophical doctrines may in all consistency reject as unargued all of my claims about the social sciences, the diagnosis offered for their alleged failures, and the prognosis made about the preemption of conventional social science by sociobiology. I do not, however, consider this an objection to the argument of this work, but a reflection of its general coherence. For I began the book by arguing that the vast issues of epistemology and metaphysics that separate the empiricist and the physicalist from the rationalist and the emergentist are not resolvable within the confines of the philosophy of science. And the issues broached within this subject are ultimately expressions of these vast questions of traditional philosophical debate. If in the end the acceptance or rejection of my argument hinges on the acceptance or rejection of empiricism and physicalism, then my argument must *perforce* be valid. Whether it is sound, whether the two doctrines that serve as my premises can ever be vindicated against their competitors, is another matter, best left to philosophers. For social scientists have more pressing concerns than the solution of the perennial questions of philosophy. If this book has successfully solved the puzzle of preserving empiricism and its methodological dicta in the face of the social scientist's failure to arrive by the employment of these dicta at results as impressive as those of natural science, then it has removed an obstacle to the prosecution of these concerns.

Notes

Chapter 1

1. John Stuart Mill, *A System of Logic* (London, 1843), bk. 6, chap. 1, sect. 2.
2. Ibid.
3. Ibid., chap. 2, sect. 1.
4. Cf. for example, Norman Malcolm, "The Conceivability of Mechanism," *Philosophical Review* 77 (1968): 45-72, and Chapter 8 below.

Chapter 2

1. John Stuart Mill, *A System of Logic* (London, 1843), bk. 6, chap. 1, sect. 2 (emphasis added).
2. In chaps. 7 and 8 of bk. 6, Mill criticizes the unwarranted importation of inappropriate methods from chemistry and geometry to the social sciences.
3. Perhaps the contemporary *locus classicus* of this view is in Peter Winch, *The Idea of a Social Science* (London: Routledge & Kegan Paul, 1958). This work appeared in what Donald Davidson has called a "tide of little red books" on "philosophical psychology," many of which argue for the same claim. These writers are heavily influenced by the later works of Ludwig Wittgenstein, and constitute the most unified set of opponents to views like Mill's and his followers' in the English-speaking world. Among Continental philosophers and social scientists, opposition to Mill's views are inspired by a Hegalian and Marxist tradition, which in recent years has shown signs of sympathy to the Wittgensteinian approach.
4. The ablest exposition of this view is to be found in Charles Taylor, *The Explanation of Behaviour* (London: Routledge & Kegan Paul, 1964). Taylor's argument is divided into two parts. In the first he purports to show that if behavior is teleological and, more specifically, intentional, then it cannot be explained in ways characteristic of natural science. His account of these ways is in general agreement with the empiricist treatment of the nature of natural science. The second part of his argument is an attempt to show that behavior even at the level of lower mammals can only be teleological or purposive. Accordingly, it cannot be accounted for by mechanistic principles or causal laws. The chief virtue of Taylor's work is that he recognizes that the issue of whether behavior is irreducibly teleological is an empirical one, and he canvasses a large number of experimental failures in order to show that as a matter of contingent fact mammalian behavior is purposive and even intentional.

5. Thus, in *Philosophical Investigations* (Oxford: Blackwell's, 1953), p. 232, Ludwig Wittgenstein writes: "The confusion and barrenness of psychology is not to be explained by calling it a 'young science'; its state is not comparable with that of physics, for instance, in its beginnings. . . . For in psychology there are experimental methods and conceptual confusions. . . . The existence of experimental methods makes us think we have the means of solving the problems which trouble us; though problem and method pass one another by." And in the preface to *Explanation and Human Action* (Berkeley: University of California Press, 1966), A. R. Louch writes: "If the social sciences are added to [this] motto," Louch's "aim as well as his indebtedness will be sufficiently indicated."

6. Cf. Winch, *Idea of a Social Science*, p. 1, and Louch, *Explanation and Human Action*, p. 2, for explicit expressions of this ridicule.

7. In speaking of the followers of Mill, I intend to cast a wide net, and include many thinkers who do not consider themselves exponents of any distinctive view associated with Mill. The persons I have in mind, in speaking of Mill's followers, have in common a commitment to empiricism and a sympathy for positivist and postpositivist thought in the philosophy of science. This will include scientific or physical realists, certain writers who characterize themselves as pragmatists, and philosophers like W.V.O. Quine and his protégés. Most of the these philosophers may well have a low opinion of Mill's work, but nevertheless would have to admit to sharing his sympathies if only they paid his writings the attention they deserve. Mill is today one of the most unjustifiably neglected English philosophers.

8. In the *System of Logic*, bk. 6, chap. 7, sect. 1, Mill writes: "Men are not when brought together converted into another kind of substance, with different properties. . . . Human beings in society have no properties but those which are derived from and may be resolved into the laws of the nature of individual man." But he also notes that "whether organic causes exercise a direct influence over [all] classes of mental phenomena is hitherto far from being ascertained" (bk. 6, chap. 4, sect. 4).

9. For example, in *Rational Economic Man* (Cambridge: Cambridge University Press, 1975), Martin Hollis and Edward Nell trace all the defects of neoclassical economics to alleged deficiencies in logical positivism, and find the origin of these, in turn, in empiricism. By contrast, they adopt a Marxian economic theory which they explicitly justify by founding its truth on that of epistemological rationalism. Presumably, their confidence in rationalistic foundations for synthetic claims extends to the natural sciences as well.

10. Good examples of such writers are R. Harré and P. F. Secord in *The Explanation of Social Behavior* (Oxford: Blackwell's, 1972). They do not recognize that their commitment to a nonempiricist account of causal powers and the nature of essential dispositions, both in the subjects of natural and social science, commits them to providing a rationalist epistemology.

11. For example, William Dray, *Law and Explanation in History* (Oxford: Oxford University Press, Clarendon Press, 1957), expounds an anti-empiricist account of causal explanation without admitting any wider philosophical ramifications.

12. Followers of the later Wittgenstein hold this view by and large.

13. Cf., for example, Winch, *Idea of a Social Science*, p. 78, and R. S. Peters, *The Concept of Motivation* (London: Routledge & Kegan Paul, 1958), p. 12.

14. Sustained denials of the appropriateness of a causal reading of reasons are to be found in Taylor, *Explanation of Behaviour*, chaps. 1-2, and in A. I. Meldin, *Free Action* (London: Routledge & Kegan Paul, 1967), p. 87ff.

15. This view animates my *Microeconomic Laws* (Pittsburgh: University of Pittsburgh Press, 1976), and it is explicitly argued for in the preface to that work.

Chapter 3

1. Emile Durkheim, *Suicide* (New York: Macmillan, 1949), pp. 38-39.

2. Ibid., p. 149.

3. Ibid., p. 151.

4. Alasdair MacIntyre, "The Idea of a Social Science," in Alan Ryan, ed., *Philosophy of Social Explanation* (Oxford: Oxford University Press, 1975), p. 27. Cf. Peter Winch, *The Idea of a Social Science* (London: Routledge & Kegan Paul, 1958).

5. Durkheim, *Suicide*, p. 44.

6. MacIntyre, "Idea of a Social Science," pp. 26-27.

7. Durkheim, *Suicide*, p. 151.

8. Emil Durkheim, *Rules of the Sociological Method* (Paris, 1901), p. 128.

9. George Homans and David Schneider, *Marriage, Authority and Final Causes* (Glencoe, Ill.: Free Press, 1955).

10. Rodney Needham, *Structure and Sentiment* (Chicago: University of Chicago Press, 1962).

11. Claude Lévi-Strauss, *Les Structures élémentaires de la parenté* (Paris, 1949), p. 558.

12. Needham, *Structure and Sentiment*, p. 28. Page references to this work are given in the text for the remainder of this chapter.

13. Homans and Schneider, *Marriage, Authority and Final Causes*, p. 26.

14. Durkheim, *Rules of the Sociological Method*, p. 133.

15. Lévi-Strauss, *Les Structures élémentaires*, pp. 107, 108, 170-71, 175.

16. Winch, *Idea of a Social Science*, pp. 123-24.

Chapter 4

1. Cf. Norman Malcolm, "The Conceivability of Mechanism," *Philosophical Review* 77 (1968): 45-72.

2. Bertrand Russell, "On the Notion of Cause," *Mysticism and Logic* (London: Longmans, 1917).

3. For a detailed discussion of the features by virtue of which economics seems more advanced than the other social sciences, see Alexander Rosenberg, *Microeconomic Laws* (Pittsburgh: University of Pittsburgh Press, 1976), chap. 1.

4. P. H. Wicksteed, *The Common Sense of Political Economy* (London, 1910), pp. 126, 13.

5. It was by virtue of their employment of this hypothesis that the "marginalists" came to be so called.

6. John Hicks, *Value and Capital* (Oxford: Oxford University Press, 1939), pp. 34 and 11.

7. L. L. Thurstone, "The Indifference Function," *Journal of Social Psychology* 2 (1931): 139-67.

8. John Von Neumann and Oscar Morgenstern, *The Theory of Games and Economic Behavior* (Princeton: Princeton University Press, 1944).

9. These experiments are reported and discussed in Ward Edwards, "The Theory of Decision Making," *Psychological Bulletin* 51 (1954): 397ff.

10. Frank Ramsey, "Truth and Probability," *Foundations of Mathematics and Other Logical Essays* (New York: Harcourt, Brace, 1931); L. J. Savage, *Foundations of Statistics* (New York: John Wiley, 1954).

11. Experiments reported in Ward Edwards, "Behavioral Decision Theory," *Annual Review of Psychology,* 1964, p. 479.

12. Gary S. Becker, *The Economic Approach to Human Behavior* (Chicago: University of Chicago Press, 1976). The quotation is from p. 14 (emphasis in the original). Subsequent page references appear in the text.

13. Hal Varian and Alan Gibbard, "Economic Models," *Journal of Philosophy* 75 (1978): 666-67.

Chapter 5

1. Paul Churchland, "The Logical Character of Action Explanations," *Philosophical Review* 79 (1970): 215. Subsequent page references appear in the text.

2. For a useful introduction to the notion of "functional characterization," see J. A. Fodor, *Psychological Explanation* (New York: Random House, 1968), esp. chap. 3.

3. For example, in *A Materialist Theory of Mind* David Armstrong argues that statements like L, which refer to mental items, will eventually be reducible to neurophysiological laws, just as Mendel's claims about the gene are explainable in molecular genetics as claims about strings of DNA.

4. For an account of the details of these issues and their development, see Monroe Strickberger, *Genetics,* 1st ed. (New York: Macmillan, 1968), esp. chap. 25, "Genetic Fine Structure." For a useful introduction to the philosophical issues, cf. Michael Ruse, *Philosophy of Biology* (London: Hutchison, 1973), and David Hull, *Philosophy of Biological Science* (Englewood Cliffs, N.J.: Prentice-Hall, 1974).

5. The notion of a principle of systematic isomorphism and its relation to the thesis of physicalism is broached in an influential paper by Richard Brandt and Jaegwon Kim, "The Logic of the Identity Theory," *Journal of Philosophy* 65 (1967): 515-37.

6. A sustained version of this argument is to be found in Donald Davidson, "Mental Events," in L. Foster and J. Swanson, eds., *Experience and Theory* (Amherst: University of Massachusetts Press, 1970), pp. 79-101, and idem, "The Material Mind," in P. Suppes, ed., *Logic, Methodology, and Philosophy of Science* (Amsterdam: North-Holland, 1973), pp. 709-22.

7. For a detailed defense of this claim, together with an account of the reduction of cistrons to strands of DNA, see Michael Ruse, "Reduction in Genetics," in R. S. Cohen and Alex Michalos, eds., *PSA 1974* (Dordrecht: Reidel, 1976), pp. 633-51.

8. As employed in the present work this distinction was first drawn in Keith Donnellan, "Reference and Definite Description," *Philosophical Review* 75 (1966): 281-304.

Chapter 6

1. Norman Malcolm, "Knowledge of Other Mind," *Knowledge and Certainty* (Englewood Cliffs, N.J.: Prentice-Hall, 1963), pp. 135-36. The passages quoted are from Ludwig Wittgenstein, *Philosophical Investigations* (Oxford: Blackwell's, 1953), pp. 113 and 178. The emphasis is Malcolm's.

2. Cf., for example, G.E.M. Anscombe, "The First Person," in S. Guttenplan, ed., *Mind and Language* (Oxford: Oxford University Press, Clarendon Press, 1975). Recent arguments along these lines have focused on the *unexplained* ability of humans to employ indexical and quasi-indexical notions (as well as other intensional devices) which, it is argued, computers cannot employ.

3. Cf. Carl Hempel and Paul Oppenheim, "Studies in the Logic of Explanation," *Philosophy of Science* 15 (1948): 135-75, reprinted in Carl Hempel, *Aspects of Scienti-*

fic Explanation (New York: Free Press, 1965), pp. 245-95, esp. sect. 6, for the *locus classicus* of this requirement for general laws. Much work in the philosophy of science has focused on the analysis of the concept of the purely qualitative predicate, but no one has suggested that Hempel and Oppenheim's requirement that all the terms in a general law be purely qualitative needs to be abandoned. In this chapter I shall assume it to be a necessary requirement for general laws, though I will not defend the requirement.

4. Cf. J.J.C. Smart, *Philosophy and Scientific Realism* (London: Routledge & Kegan Paul, 1963).

5. Much of the argumentation of the preceding pages I owe to the influence of David Hull. See especially "Are Species Really Individuals?" *Systematic Zoology* 25 (1976): 174-91; and "A Matter of Individuality," *Philosophy of Science* 45 (1978): 335-60. The latter paper reflects conclusions about the social sciences arrived at independently of my own but highly similar to them. I owe much of my appreciation of these issues to Hull's work.

6. Alexander Rosenberg, "Genetics and the Theory of Natural Selection: Synthesis or Sustenance?" *Nature and System* 1 (1978): 3-15; and "The Supervenience of Biological Concepts," *Philosophy of Science* 45 (1978): 368-87. The presentation here and in those two papers is an informal account of the axiomatic system for the theory of natural selection constructed in an unjustly neglected paper by Mary B. Williams, "Deducing the Consequences of Evolution," *Journal of Theoretical Biology* 29 (1970): 342-85.

7. See Williams, "Consequences of Evolution," p. 380, n. 6.

8. A. J. Lotka, *Elements of Physical Biology* (Baltimore: Williams and Wilkins, 1925), and V. Volterra, "Variazione e fluttrazini del numero d'individui in specie animali conviventi," *Memoria Accademica Nazionale Lincei*, 6th ser. 2 (1926): 31-113.

9. J. T. Tanner, "The Stability and the Intrinsic Growth Rates of Prey and Predator Populations," *Ecology* 56 (1975): 855-67.

10. Cf. R. M. May, *Stability and Complexity in Model Ecosystems* (Princeton: Princeton University Press, 1973). The original proofs were provided in the 1930s.

11. Adapted from E. C. Pielou, *Mathematical Ecology*, 2nd ed. (New York: Wiley, 1977), p. 107.

12. E. L. Thorndike, *Animal Intelligence* (New York: Macmillan, 1911).

13. A popular exposition of these findings appeared in James Olds, "Pleasure Centers of the Brain," *Scientific American* 195 (February 1956): 105-16. See also his "Self-Stimulation of the Brain," *Science* 127 (1958): 315-24.

14. J. A. Fodor, *Psychological Explanation* (New York: Random House, 1968), p. 136.

15. Cf. for example, Alan Newell and H. A. Simon, "Computers in Psychology," in R. D. Luce, R. R. Bush, and E. H. Galanter, eds., *Handbook of Mathematical Methods in Psychology* (New York: John Wiley, 1963). See also Alan Newell and H. A. Simon, "GPS: A Program That Simulates Human Thought," in M.E.A. Feigenbaum and J. Feldman, eds., *Computers and Thought* (New York: McGraw-Hill, 1963), pp. 279-93.

16. W.V.O. Quine, *Word and Object* (Cambridge: MIT Press, 1960), p. 221.

Chapter 7

1. See Herbert Spencer, *The Principles of Sociology*, 3 vols. (London, 1876-96).

2. The *locus classicus* for this sort of skepticism is Carl Hempel, "The Logic of Functional Analysis," *Aspects of Scientific Explanation* (New York: Free Press, 1965).

3. Confused versions of the stronger and weaker theses of biological determinism can be found, along with their alleged political ramifications, in E. Allen et al., "Socio-

biology: Another Biological Determinism," *Bioscience* 26 (1976): 182-86, and in E. Allen et al., "Sociobiology: A New Biological Determinism," in *Biology as a Social Weapon* (Minneapolis: Burgess, 1977). A more temperate elaboration of the weak thesis, together with an attack on Wilson for allegedly embracing it, is to be found in Stephen J. Gould, *Ever Since Darwin* (New York: Norton, 1977); see, in particular, "Biological Potentiality vs. Biological Determinism," ibid., pp. 251-59.

4. E. O. Wilson, *Human Nature* (Cambridge: Harvard University Press, 1978), p. 16.

5. E. O. Wilson, *Sociobiology: The New Synthesis* (Cambridge: Harvard University Press, 1975), p. 23.

6. Ibid.

7. Ibid., p. 6 (emphasis added).

8. I have discussed the literature of philosophy's recent treatment of this problem in "Causation and Teleology in Contemporary Philosophy of Science," in the *Chronicles of Philosophy: Philosophy of Science* (Paris: International Institut de Philosophie, 1980).

9. Wilson, *Sociobiology*, p. 20.

10. Wilson, *Human Nature*, p. 32.

11. One recent example of a biologist's arguing vigorously for the tautological character of the theory of natural selection is R. H. Peters, "Tautology in Evolution and Ecology," *American Naturalist* 110 (1976): 1-12. Characteristic of his claims is the following: "The 'theory of evolution' does not make predictions, . . . but is instead a logical formula which can be used only to classify empiricisms [sic] and show the relationships which such a classification implies" (p. 1).

12. Mary B. Williams, "Deducing the Consequences of Evolution," *Journal of Theoretical Biology* 29, (1970): 359.

13. Jaegwon Kim, "Supervenience and Nomological Incommensurables," *American Philosophical Quarterly* 15 (1978): 27-36.

14. For an account of the theoretical and computational uses to which biologists put the notion of fitness, cf. W. H. Bossert and E. O. Wilson, *Primer of Population Biology* (Stamford, Conn.: Sinauer, 1971).

15. I attempt to substantiate this claim in Alexander Rosenberg, "Genetics and the Theory of Natural Selection: Synthesis or Sustenance?" *Nature and System* 1 (1978): 3-15.

16. Cf. D. C. Dennett, "Why the Law of Effect Won't Go Away," *Brainstorms* (Montgomery, Vt.: Bradford Books, 1979), for a sustained discussion of the parallel between the law of effect and the theory of natural selection.

Chapter 8

1. E. O. Wilson, *Sociobiology: The New Synthesis* (Cambridge: Harvard University Press, 1975), chap. 27.

2. Cf., for instance, Karl Popper, *The Poverty of Historicism* (London: Routledge & Kegan Paul, 1957).

3. See the discussion of Needham in Chapter 3 above.

4. David Lack, *Darwin's Finches* (Cambridge: Cambridge University Press, 1947).

5. E. C. Pielou, *Mathematical Ecology*, 2nd ed. (New York: Wiley, 1977), p. 105.

6. James Quirk and Richard Ruppert, "Qualitative Economics and the Stability of Equilibrium," *Review of Economic Studies* 32 (1965): 311-26.

7. J. Maynard Smith, *Models in Ecology* (Cambridge: Cambridge University Press, 1974), pp. 121ff., 59-61.

8. Gary S. Becker, *The Economic Approach to Human Behavior* (Chicago: University of Chicago Press, 1976), p. 284.

9. Wilson, *Sociobiology*, p. 3.

10. E. O. Wilson, *Human Nature* (Cambridge: Harvard University Press, 1978), p. 154.

11. Cf. the discussion of Durkheim's reasons for agnosticism on this score in Chapter 3 above.

12. Wilson, *Sociobiology*, p. 118.

13. Wilson, *Human Nature*, p. 219.

14. Marshall Sahlins, *The Use and Abuse of Biology* (Ann Arbor: University of Michigan Press, 1976). Subsequent page references appear in the text.

15. Cf. the quotation documented by n. 10 above.

16. Wilson, *Sociobiology*, p. 550.

17. Ibid., p. 554.

18. Wilson, *Human Nature*, chap. 7, and esp. p. 157.

19. Quoted by Rodney Needham in *Structure and Sentiment* (Chicago: University of Chicago Press, 1962), p. 126, from Emil Durkheim's *Rules of the Sociological Method* (Paris, 1901), p. 102.

20. Needham, *Structure and Sentiment*, p. 73.

21. Peter Winch, *The Idea of a Social Science* (London: Routledge & Kegan Paul, 1958).

22. Norman Malcolm, "The Conceivability of Mechanism," *Philosophical Review* 77 (1968): 45-72. Subsequent references appear in the text.

23. Keith Donnellan, "Reference and Definite Description," *Philosophical Review* 75 (1966): 281-304.

24. Cf. Ludwig Wittgenstein, *Philosophical Investigations* (Oxford: Blackwell's, 1956), p. 232, and expositions of this claim in Winch, *Idea of a Social Science*, n. 21; A. R. Louch, *Explanation and Human Action* (Berkeley: University of California Press, 1969); R. S. Peters, *The Concept of Motivation* (London: Routledge & Kegan Paul, 1958); and A. I. Meldin, *Free Action* (London: Routledge & Kegan Paul, 1967).

Bibliography

Allen, E., et al. "Sociobiology: Another Biological Determinism." *Bioscience* 26 (1976): 182-86.

_____. "Sociobiology: A New Biological Determinism." In *Biology as a Social Weapon*. Minneapolis: Burgess, 1977.

Anscombe, G.E.M. "The First Person." In S. Guttenplan, ed., *Mind and Language*. Oxford: Oxford University Press, Clarendon Press, 1975.

Armstrong, D. *A Materialist Theory of Mind*. London: Routledge & Kegan Paul, 1964.

Becker, G. *The Economic Approach to Human Behavior*. Chicago: University of Chicago Press, 1976.

Bossert, W. H., and Wilson, E. O. *Primer of Population Biology*. Stamford, Conn.: Sinauer, 1971.

Brandt, R., and Kim, J. "The Logic of the Identity Theory." *Journal of Philosophy* 65 (1967): 515-37.

Churchland, P. "The Logical Character of Action Explanations." *Philosophical Review* 79 (1970): 214-36.

Davidson, D. "Mental Events." In L. Foster and J. Swanson, eds., *Experience and Theory*. Amherst: University of Massachusetts Press, 1970.

_____. "The Material Mind." In P. Suppes, ed., *Logic, Methodology and Philosophy of Science*, pp. 709-22. Amsterdam: North-Holland, 1973.

Dennett, D. C. *Brainstorms*. Montgomery, Vt.: Bradford Books, 1979.

Donnellan, K. "Reference and Definite Description." *Philosophical Review* 75 (1966): 281-304.

Dray, W. *Law and Explanation in History*. Oxford: Oxford University Press, Clarendon Press, 1957.

Durkheim, E. *Rules of the Sociological Method*. Paris, 1901.

_____. *Suicide*. New York: Macmillan, 1949.

Edward, W. "The Theory of Decision Making." *Psychological Bulletin* 51 (1954): 380-417.

_____. "Behavioral Decision Theory." *Annual Review of Psychology* 12 (1964): 473-98.

Feigenbaum, M.E.A., and Feldman, J., eds. *Computers and Thought*. New York: McGraw-Hill, 1963.

219

Fodor, J. A. *Psychological Explanation*. New York: Random House, 1968.

Gould, S. J. *Ever Since Darwin*. New York: Norton, 1977.

Harré, R., and Secord, E. *The Explanation of Social Behavior*. Oxford: Blackwell's, 1972.

Hempel, C. G. *Aspects of Scientific Explanation*. New York: Free Press, 1965.

Hicks, J. *Value and Capital*. Oxford: Oxford University Press, 1939.

Hollis, M., and Nell, E. *Rational Economic Man*. Cambridge: Cambridge University Press, 1975.

Homans, G., and Schneider, E. *Marriage, Authority and Final Causes*. Glencoe, Ill.: Free Press, 1955.

Hull, D. *Philosophy of Biological Science*. Englewood Cliffs, N.J.: Prentice-Hall, 1974.

————. "Are Species Really Individuals?" *Systematic Zoology* 25 (1976): 174-91.

————. "A Matter of Individuality." *Philosophy of Science* 45 (1978): 335-60.

Lack, D. *Darwin's Finches*. Cambridge: Cambridge University Press, 1947.

Lévi-Strauss, C. *Les structures élémentaires de la parenté*. Paris, 1949.

Lotka, A. J. *Elements of Physical Biology*. Baltimore: Williams and Wilkins, 1925.

Louch, A. R. *Explanation and Human Action*. Berkeley: University of California Press, 1966.

Luce, R.; Bush, R.; and Galanter, E., eds. *Handbook of Mathematical Methods in Psychology*. New York: John Wiley, 1963.

MacIntyre, A. "The Idea of a Social Science." In A. Ryan, ed., *Philosophy of Social Explanation*. Oxford: Oxford University Press, 1975.

Malcolm, N. *Knowledge and Certainty*. Englewood Cliffs, N.J.: Prentice-Hall, 1963.

————. "The Conceivability of Mechanism." *Philosophical Review* 77 (1968): 45-72.

May, R. M. *Stability and Complexity in Model Ecosystems*. Princeton: Princeton University Press, 1973.

Mill, J. S. *System of Logic*. London, 1843.

Needham, R. *Structure and Sentiment*. Chicago: University of Chicago Press, 1962.

Olds, J. "Pleasure Centers of the Brain." *Scientific American* 195 (February 1956): 105-16.

————. "Self-Stimulation of the Brain." *Science* 127 (1958): 315-24.

Peters, R. H. "Tautology in Evolution and Ecology." *American Naturalist* 110 (1976): 1-12.

Peters, R. S. *The Concept of Motivation*. London: Routledge & Kegan Paul, 1958.

Pielou, E. C. *Mathematical Ecology*. 2nd ed. New York: Wiley, 1977.

Popper, K. *The Poverty of Historicism*. London: Routledge & Kegan Paul, 1957.

Quine, W.V.O. *Word and Object*. Cambridge: MIT Press, 1960.

Quirk, J., and Ruppert, R. "Qualitative Economics and the Stability of Equilibrium." *Review of Economic Studies* 32 (1965): 311-26.

Ramsey, F. *Foundations of Mathematics and Other Logical Essays*. New York: Harcourt, Brace, 1931.

Rosenberg, A. *Microeconomic Laws: A Philosophical Analysis*. Pittsburgh: University of Pittsburgh Press, 1976.

_____. "Genetics and the Theory of Natural Selection: Synthesis or Sustenance?" *Nature and System* 1 (1978): 3-15.

_____. "The Supervenience of Biological Concepts." *Philosophy of Science* 45 (1978): 368-87.

_____. "Causation and Teleology in Contemporary Philosophy of Science. In *Chronicles of Philosophy: Philosophy of Science*. Paris: International Institut de Philosophie, 1980.

Ruse, M. *Philosophy of Biology*. London: Hutchison, 1973.

_____. "Reduction in Genetics." In R. S. Cohen and A. Michalos, eds., *PSA 1974*. Dordrecht: Reidel, 1974.

Russell, B. *Mysticism and Logic*. London: Longmans, 1917.

Sahlins, M. *The Use and Abuse of Biology*. Ann Arbor: University of Michigan Press, 1976.

Smart, J.J.C. *Philosophy and Scientific Realism*. London: Routledge & Kegan Paul, 1963.

Smith, J. M. *Models in Ecology*. Cambridge: Cambridge University Press, 1974.

Spencer, H. *The Principles of Sociology*. 3 vols. London, 1876-96.

Strickberger, M. *Genetics*. New York: Macmillan, 1968.

Tanner, J. T. "The Stability and the Intrinsic Growth Rates of Prey and Predator Populations." *Ecology* 56 (1975): 855-67.

Taylor, C. *The Explanation of Behaviour*. London: Routledge & Kegan Paul, 1964.

Thorndike, E. L. *Animal Intelligence*. New York: Macmillan, 1911.

Thurstone, L. L. "The Indifference Function." *Journal of Social Psychology* 2 (1931): 139-67.

Varian, H., and Gibbard, A. "Economic Models." *Journal of Philosophy* 75 (1978): 666-77).

Volterra, V. "Variazione e fluttrazini del numero d'individui in specie animali conviventi." *Memoria Academica Nationale Lincei*, 6th ser. 2 (1926): 31-113.

Von Neumann, J., and Morgenstern, O. *The Theory of Games and Economic Behavior*. Princeton: Princeton University Press, 1944.

Wicksteed, P. H. *The Common Sense of Political Economy*. London, 1910.

Williams, M. B. "Deducing the Consequences of Evolution." *Journal of Theoretical Biology* 29 (1970): 342-85.

Wilson, E. O. *Sociobiology: The New Synthesis*. Cambridge: Harvard University Press, 1975.

_____. *On Human Nature*. Cambridge: Harvard University Press, 1978.

Winch, P. *The Idea of a Social Science*. London: Routledge & Kegan Paul, 1958.

Wittgenstein, L. *Philosophical Investigations*. Oxford: Blackwell's, 1953.

Index